Manufacturing in Real Time

Managers, Engineers, and an Age of Smart Machines

Dedication

This book is dedicated to the unsung heroes who make airplanes, CAT scanners, cereal packages, washing machines, and all the other things we use in our daily life.

Manufacturing in Real Time

Managers, Engineers, and an Age of Smart Machines

Gian F. Frontini, Scott L. Kennedy

An imprint of Elsevier Science
www.bh.com

Amsterdam, Boston, London, New York, Oxford, Paris, San Diego, San Francisco, Singapore, Sydney, Tokyo

Butterworth-Heinemann is an imprint of Elsevier Science.

Copyright © 2003, Elsevier Science (USA). All rights reserved.

 Recognizing the importance of preserving what has been written, Elsevier Science prints its books on acid-free paper whenever possible.

Library of Congress Cataloging-in-Publication Data

0-7506-7722-8

British Library Cataloguing-in-Publication Data

A catalogue record for this book is available from the British Library.

The publisher offers special discounts on bulk orders of this book.
For information, please contact:

Manager of Special Sales
Elsevier Science
200 Wheeler Road
Burlington, MA 01803
Tel: 781-313-4700
Fax: 781-313-4882

For information on all Butterworth-Heinemann publications available, contact our World Wide Web home page at: http://www.bh.com

10 9 8 7 6 5 4 3 2 1

Printed in the United States of America

Table of Contents

Introduction xi

Chapter 1
The Historical Development of Flow in Manufacturing 1
    Manufacturing: Definition and Origin 2
    Evolution of Manufacturing and the Emergence of Flow . . 3

Chapter 2
Dynamics in the Marketplace 19
    Introduction to Supply Chain Management 20
    The Technical Nature of ISCM 24
    The Strategic Role of ISCM 25
    Customer Value 26
    Forces in the Marketplace 30

Chapter 3
Introduction to Product Streams 39
    What Is a Product Stream? 40
    Mass Flow 42
    Cash Flow 48
    Asset Cost 50
    Information Flow 53
    Flow, Yield, and Product Costing 54

Chapter 4
Automation in the Factory 57
    The Role of Automation in the Supply Chain of
      Manufactured Products 58
    Talking to the Machines 60
    The Self-Taught Machine 62
    Feed-Forward, Feedback, and Adaptation 64
    From Cogs to Networks 66
    Analog and Digital Controllers 69
    How Machines Learn 72
    The Intelligent Machine and Real People 78

Chapter 5
Variation in Dynamic Systems 81
    Variation in Static Systems 82
    Manufacturing Industry and the Use of Statistical
        Process Control in Dynamic Systems 84
    Statistical Distributions and the Measurement
        of Variation 88
    Shewhart's Control Charts 94
    The Random Walk of Manufacturing Processes 96

Chapter 6
The Statistical Meaning of Profitable Manufacturing 99
    Managers, Engineers, and Statistical Methods 100
    Measuring Variation in Product Streams 101
    What Is Profitable Manufacturing? 107

Chapter 7
In Praise of Simplicity 111
    The Perplexing Green Lawn 112
    An Approach to Simple Design 114

Chapter 8
Organizations for Real-Time Decision Making 123
    The Importance of Decision Making 124
    Organizational Structures 126
    Rewards and Challenges 136
    On Human Frailty 138
    Glimpses of the Future 141

Chapter 9
Next-Generation Manufacturing and the Virtual
   Reality Plant 143
    The Next-Generation Manufacturing Project 144
    Redesigning the Manufacturing Enterprise 146
    Understanding the Gap 149
    Building a Dynamic Business Model to Improve
        Decision Making 156

The Virtual Factory:
   Designing Product Streams with Dynamic Models ... 159

Chapter 10
Industrial Engineers and Intelligent Machines 163
   Sizing the Gaps 164
   The Pressure to Comply 168
   Man, Machines, and Product Flow 171
   The Global Manufacturing Business and the
      Measure of Productivity 172
   People, Machines, and the Workplace 175
   The Preextinction Society 178
   Dealing with Globality 181
   The New Generation of Industrial Engineers 182

References 183

Glossary 191

Index 203

Introduction

This book is about making things useful in everyday life and about making them *well*.

There is no limitation to the types of objects we buy and use, either directly or indirectly. We use bathtubs, washing machines, cereal packages, and televisions. Other objects, such as electron microscopes or CAT scanners, we use through services we acquire from a laboratory or a hospital. Most of the time, we do not stop to think about the origin of these objects; we take their existence for granted as part of our society. These objects have one thing in common: somebody has designed and manufactured them *well*.

Think of the meaning of the word "well" as applied to manufacturing; it means a lot more than a good product.

"Well" means giving the consumer value for the money he/she pays. "Well" means making an adequate profit at each step of the long road from the raw materials to the finished product. "Well" means using a minimum of resources in making the product and planning to reuse its recyclable content. "Well" also means pride and satisfaction in a job well done.

Achieving all these goals by designing reliable manufacturing systems and business processes is the essence of industrial engineering in an age of smart machines.

In the 1920s, the first industrial engineers measured the efficiency of labor in mass production plants; in this endeavor, they laid the foundations of modern manufacturing and developed the measures of industrial success.

At about the same time Shewhart and Deming created statistical process control and the principles of manufactured product quality. It took 60 years for these techniques to become accepted in the industrial world and to impact on the value of consumer products. Some say these early pioneers were visionaries ahead of their time; some say they lacked the tools for implementation and could not handle the mathematical complexity of statistical control with slide rulers. Both statements are probably true.

It is also true that in the past 40 years major changes in our society have been brought about by the development of communications and the global market. If consumers did not know what to expect from a product in the 1920s, they certainly do now, and they also have the choice of alternative sources at their disposal.

Communication infrastructures and smart machines have come at the same time and are related to each other in their ability to span systems that cover entire plants, cities, and continents. Slowly at first, with computer-assisted elevators, which could stop within a few millimeters of the floor level after impressive accelerations and decelerations and with numerically controlled machining shops, automation has become pervasive in our society.

How far are the smart machines going to develop? The pace of connectivity and functional distribution is ever increasing and the patterns of machine/communication constructs are increasingly mimicking organic systems. This is not surprising since we have created machines, communications, and business processes, and we are biological organisms.

The drive for continuing development of automation is inherent in the type of products used in our technological society. Economic justification of automation is no longer replacing labor with less expensive machines, as it was during the development of mass production. Instead, it is making things with a level of accuracy and consistency humans cannot match. The need to conserve resources and protect our environment is adding new incentives; only smart machines will be able to control processes with many times the efficiencies of those we use today. It is only automation and global communications that will make it possible to sustain more than six billion people on this planet. Returning to nature is no longer an option.

The industrial engineers are now in a new role in manufacturing businesses. They have become the designers of supply chains: a much broader task than factory-based logistic and efficiency programs. In high-tech industries, the role is recognized as essential to the business process. Large manufacturing is also discovering that it is the design of the supply chain that determines the business process and its economic viability. This is not a task that business leaders can abdicate to engineers; it is the essence of manufacturing management.

Understanding a product supply chain is a difficult task since a simple static description is insufficient to portray production and economic capability. The ideas we are describing in this book are based on a dynamic view of the business process, driven by variation as well as design. Variation is inherent in all dynamic systems (and business is no exception), but far from being a negative, variation is the key to evolution toward better performance.

The idea of exploiting variation to advantage is the theme of the research on dynamic modeling of advanced manufacturing systems done at the Centre for Automotive Materials and Manufacturing. This book describes some of the concepts in accessible language and simple mathematics.

The book also addresses the impact of new technologies on the future shape of the manufacturing industry. New technologies and new ways of doing business require changes in organizations, skills, rewards, and recognition. The level of complexity in global supply chains can no longer be controlled by a few bright individuals and this places increasing importance on skills, education, and self-initiative at the action points of the business.

We have just started this transition in the past few years and the popular press still portrays manufacturing as the domain of eccentric business leaders as it was in the early days of the Industrial Revolution. There are no dramatizations, so far as we are aware, of engineers and operating technicians, the people who are creating the manufacturing businesses of the future. The subject may be unappealing because of the lack of nudity, profanity, and violence in a safety-conscious manufacturing environment.

More likely, the world is still unaware of the social transitions that we will all experience in dealing with smart machines.

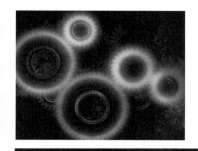

Chapter 1

The Historical Development
of Flow in Manufacturing

Manufacturing: Definition and Origin

The instinctive human desire to make things, whether useful or not, has been an important part of our culture since the dawn of time. Not much has changed since then. Humankind has made great progress in reducing the physical labor involved in the creation of goods, and the things we are making are increasingly abstract. However, the recipe for a world free from toil is still eluding us.

The focus of the postindustrial society may now have shifted to the softer sides of business, such as service, communication, education, and social care; over 60% of North American Gross Domestic Product (GDP) is in this category. This simply means that it is easier for us to manufacture things with the help of machines, so that we can dedicate our minds to other endeavors. We still need real cars, toasters, and bathtubs. Not only does somebody have to make these, they must make them well. We have to acknowledge that the survival of mankind is forever tied to making things at the expense of personal effort and manufacturing is here to stay. This book is about manufacturing and the need to do it efficiently and with a minimum of waste.

Manufacturing is defined in the *Encyclopedia Britannica*[1] as:

> "An industry that makes products from raw materials by the use of manual labor or machinery, and that is usually carried out systematically with a division of labor."

It would appear that the world at large still sees manufacturing as the mass production factories of the 1920s depicted in the movie *Modern Times* starring Charlie Chaplin. The world of business has made huge steps away from this approach in the past 30 years, but clearly it has not demonstrated the change to the general public. We are at the beginning of a new stage in the evolution in manufacturing. A better understanding of its potential may show that manufacturing businesses, including their competitive practices, are at the core of improving our chances of survival on this planet for a little longer.

Improvement driven by competition is not a new evil of industrialized society; it is one of the traits of our species and shows an ability to adapt. You can almost hear the caveman of years ago, swinging the new tool he fashioned, saying to himself:

> "If I can carve more stones faster than my neighbor, by the use of this wonderful new hammer, then I will be able to gather more food and skins, and maybe have a few tools left over to trade for some new wives, and maybe become chief of the band, and maybe conquer."

This book is about matching competitive practices with resource-efficient manufacturing.

Evolution of Manufacturing and the Emergence of Flow

Craft Production

Until the late eighteenth century, manufacturing was seen as a service industry separate from other aspects of society. The state rulers and the church controlled agriculture, financial services, communications, education, health, and welfare. The creation of goods was a necessity, but not an integral part of the ruling society. Craft production was the standard in these times. Craftsmen fashioned a complete object from the time it was a pile of raw materials until it was sold to the public as a finished item, often with a high degree of customization. For example, the hilt of a sword could be crafted to fit the hand of the owner and the owner's family crest could be engraved on the blade.

Craftsmen would apprentice with an established master for years to learn a given trade.[2] They often produced their own tools, were identified with their particular craft, and perceived their contribution to society in terms of the quality of the object they produced.[3] Hence, the contemporary connotation of craftsmanship as pertaining to skill and creativity.

There is some evidence that as early as the sixteenth century, certain trades had conceived the idea of organized shop flow. The bookbinding trade is one example. The term "forwarding" refers to the preparing of the text block by blocking it, sewing it on cords, rounding it, and attaching it to boards.[4] Another set of craftsmen performed the "finishing" operation, which involved decorating, polishing, adding hardware, and titling. In bookbinding, as in many other manufacturing operations, the construction of the product involves certain sequences. In traditional bookbinding, it is not possible to "round" the back of a book before sewing it, nor is it possible to attach boards before "rounding" since the book would not have a "shoulder" on which to fit the hinge. These product-related sequences stimulated the idea of "flow" and ways to improve the efficiency of the operations and the quality of the product by applying the best skills to each step of the operation.

In many ways the bookbinding shops of the sixteenth century had discovered the modern concept of lean manufacturing on a small scale and were able to combine high quality with productivity. The concept of yield was introduced at the same time. Books were still very scarce and paper in limited supply. Errors in binding were not acceptable and yield had to be 100%.[i] Bindings of this period and the following two centuries have survived in excellent condition, providing proof of the quality of the original product (particularly when compared to the disintegrating leather bindings of the Victorian period).

---

i. The consequences of error for an individual were swift and brutal punishment, a method not favored in modern agile manufacturing, but errors were rare since apprenticeship for bookbinding took 12 years.

The Industrial Revolution

The shift away from craft production began in England in the late eighteenth century. The Industrial Revolution marked the beginning of a society where manufacturing was integrated into other aspects of life. The same movement took place in other countries during the following centuries, but we will limit our discussion to the major shifts that took place in England, as they are representative of later events in other countries.

Early industrialists discovered the engineering application of flow as a tool to fragment manufacturing into a series of unskilled tasks with great initial results.

In the eighteenth century there were, as there are today, two different drivers for the introduction of machinery, both based on efficiency principles:

- Substitution of manual operations with machines, with the objective of attaining accuracy and repeatability beyond human capability. (This includes the case of metal screws and bolts, which only became available in quantities with the advent of machines capable of tracing helices.)[5]

- Substitution of manual operations with the objective of increasing productivity and reducing product cost. This still applies to many primary products in chemical, mining, and agricultural operations.

Inventions in the cotton industry (Kay's flying shuttle of 1733, Hargreaves' spinning jenny of 1764, Arkwright's water frame of 1769, Crompton's spinning mule of 1779, and Cartwright's power loom of 1785)[6] brought the machine into the manufacturing world. New energy sources such as steam, electricity, petroleum, and the internal-combustion engine were used to drive the machines and helped propel us toward an age of even greater automation.[7]

Material Flow

In 1784 Oliver Evans created the first production line by sequencing five machines in his grain factory and automating the movement through the machines. This is the first evidence of organized factory flow that employs the principles of machine specialization and automated materials movement.[ii] The idea of the production line is significant because by timing material transfer through conveyors, the line addresses productivity of the overall shop rather than that of a single machine.

Early factories were focused on products, not on processes, and in this sense were quite similar to many modern manufacturing systems. The formal industrial engineering discipline of process optimization had not yet been invented and factories had a high degree of flexibility in both the type of product and the output. Even in this industrialized production method there was much pride in the quality of the product, and pleasing the customer was still seen as an essential part of business success. Industrially produced goods became great competition for craft-produced goods. Some early nineteenth century guides to gentlemen gardeners state that commercially built hothouses, for growing exotic plants, far exceeded those produced by the local craftsmen in terms of quality and provided many elegant cast-iron designs of cornices at a moderate price.

Changing Skill Requirements

With the introduction of this new manufacturing system, the skill profile of the worker changed. Suddenly each worker had only one specific task requiring one specific skill (which was often acquired onsite). The physical effort involved was also dramatically reduced. This opened the factories to workers of varying ages and abilities. Gone were the multiskilled craftsmen of the preindustrial period; women and children became an important source of labor.[iii]

---

ii. The word automated is used here in a generic sense, referring to the performing of a task through a machine using power other than that of the human body, not in the specialized sense used by control engineers.

iii. The development of factories and the boom in the manufacturing industry led to the exploitation of young children as workers, an exploitation that was finally stopped in 1833 with the Factory Act.[8]

Early industrial production also sowed the seeds of dissent. Expectations for a better quality of life were at an all-time high for a large segment of the population, while the distribution of wealth overwhelmingly favored the factory owners.

Population and Demographics

Productivity in the agricultural industry increased dramatically with the invention of new implements and machines that allowed for more intense cultivation.[9] Most importantly, from a demographic viewpoint, new technologies increased sanitation through the separation of drinking water from waste by the use of steel or iron pipes. The population increased and the growing number of people needed places to live and work. The new factories were a draw for people of all backgrounds and abilities. Many farmers, unable to afford the technological advances in the agricul-tural industry, were forced into the growing urban centers to work in the factories. Cities emerged, as these farmers and others crowded to the areas encompassing factories. Urban centers, roads, and railways altered England's landscape with faster and faster means of transferring goods: flow of materials was no longer enough, it had to be fast flow.

Economics and Politics

At the beginning of the Industrial Revolution, the few factory owners were becoming wealthy while their workers struggled to survive on their meager wages (and often those of their whole family!). However, as the Industrial Revolution unfolded, manufacturing profits became more widely distributed and land declined in importance as a source of wealth.[10] This allowed the working class to become larger, more powerful, and wealthier, forcing changes upon the authority figures of the state and church, many of which had enjoyed complete rule over European countries until this point. The most obvious and sudden of these changes was the French Revolution of the late eighteenth century, in which the emerging "middle class" overthrew the monarchy in a bloody battle that allowed creative leaders to emerge and build business in a form similar to what we know today.

Mass Production

With the introduction of technical advancements during the Industrial Revolution, the drive toward improvement of the economic performance of factories led to an important discovery: material flow through a factory could be increased if both operations and product components were standardized. This new system of manufacturing became known as "mass production." In addition to the principles of specialization and division of labor, mass production introduced the standardization of parts to the manufacturing of goods.[11]

Mass production focuses on process as opposed to product. It is the methods and the components used in creating the product that are standardized, not the product itself. In its earlier form, mass production allowed for a variety of products to be made from standardized components. In this way, it was able to overtake handcrafted production. The products were of the same quality or higher. They were cheaper, more readily available, and fashionable.

Given these advantages, mass production was rapidly adopted into other countries, and by the middle of the nineteenth century large factories could be found in much of the world.[12]

Changing Markets

The Industrial Revolution of the late eighteenth century did not create new markets immediately. The first step was the centralization of manufacturing, where craft manufacturers were put out of business one after the other. This process progressed for almost a century until it reached the point where mass-produced goods were the norm and handcrafted articles the exception.

Over time, the role of craft manufacturing was forced to change. A few business-oriented craftsmen set themselves as leaders in design, fashion, and other niches, creating the modern roles of product developer and design consultant. They became an essential part of a world whose forces were pulling toward mass production.

Human faith and confidence in mass production accelerated throughout the nineteenth century and the early part of the twentieth century. Canals gave way to railways, sailing ships to steamers, stone to steel, candles to gas and kerosene - all of them cheaper, more efficient, and more practical systems!

Ford

In the early twentieth century, Henry Ford and Alfred Sloan improved and expanded the system of mass production to introduce the concept into the automotive industry. Ford revolutionized the automotive industry and outdated the horse and buggy as the common form of transportation by producing motor vehicles affordable to the masses. This could never have been done without mass production. In 1923 a Model T roadster sold for $360, a price reduced to combat the recession.[13] Compare this to the average Ford factory-worker wage of $5 per day at the same time (with no tax deductions) and Ford had achieved incredible market appeal, with fewer than 80 work days needed by a Ford worker to have enough money to buy the Model T.[14]

For the sake of argument, we will compare the Model T to today's Ford Focus (please understand this is a loose comparison only). A Ford Focus costs approximately $14,000.[15] With industrial factory wages in Detroit of $16.87 per hour, or approximately $135 per day,[16] a Focus would take approximately 20 more days of work to buy as the Model T did in 1923. Should we decide to measure industrial productivity only by customer purchasing power, the argument can be made that Ford in 1923 was at a slightly higher level than Ford is today, after 80 years of improvements!

Many manufacturing businesses are rediscovering the science and the wisdom Ford built into his business. Ford's vision included all the imperatives of customer value built into modern manufacturing, as well as the economic compromises required to create a mass market for mass-produced goods. The model T was black, but you could paint it any color you wanted for a very small price in a local shop and add numerous accessories from aftermarket parts manufacturers. It is hard to find another product of its era with so much value built into it.

Ford's thinking was a product of technical methodology applied to manufacturing. An examination of Ford's business shows a logical construction of supply chains, as we are rediscovering today. The weak link in the 1920 supply chain was the flow of information, and supply chains could not be sustained without a strong management directive and tremendous personal effort. The new science of industrial engineering was to change that.

The Birth of Industrial Engineering

The early part of the twentieth century also saw the efforts of pioneer consultants, such as Frederic Taylor, Frank Gilbreth, and others, lay the seeds of a more systematic approach to the design and management of mass production.[17] Operating in academia and as independent professionals, Taylor and his peers created rules for the measurement of industrial times and for the definition of time-related productivity. This had profound impacts, from dramatic improvements in factory design, task distribution, and productivity to the restructuring of a company's hierarchical management.[18] The modern methodology of lean manufacturing and statistical process control owes much to these early pioneers of industrial engineering.

Taylor's views are not fashionable today because of his rigid style.[iv] Still, they were revolutionary in the context of his society and the social conditions prevailing in the factories of the time. Industrial engineering did help the workers, because it brought about more equitable measures in determining wages and working conditions and at the same time produced improved earnings for the business. The industrial engineering approach survived the Great Depression and the labor unrest of the following 20 years. It was also instrumental in creating industrial powers in most developing countries.

---

iv. Taylor's rigid style implies that all human and machine-run activities could be encompassed in mathematical relationships. Today, we would add "soft" issues to this, which are not so easily quantifiable.

Manufacturing became so important that the years between the wars saw belief in the machine elevated to a cult. The huge turbines of the hydro generating plants and the great locomotives, so wonderfully complex and elegant, were objects to be admired and revered as they made man look so small and smart! Early automation efforts based on hydraulics and mechanical linkages compounded this idea of the "supermen-masters of the world," tapping unlimited resources and feeding them to factories with fewer and fewer workers.

The early industrial engineers and designers of automated factories inadvertently sowed an aggressive seed in the midst of a confident industrial society: the importance of the production process was above everything else. This seed was to germinate during World War II and the postwar industrial revival and render U.S. and European businesses unprepared for global competition. Great machines and efficient processes existed, but many of them were producing products customers did not value.[v]

The idea of optimizing the production process was excellent. It was the obsessive application of the idea, and in particular its use to avoid the difficulties of managing supply chains involving customers and suppliers, that created huge industrial bureaucracies. This led western industry into a 40-year decline that began during WWII and intensified in the postwar industrial revival.[vi]

The Japanese Approach

During World War II Japan found itself forced into major changes, some of which led to the evolution of a new manufacturing philosophy. Because it was difficult to get outside suppliers for the electrical components of their vehicles, Toyota started to make their own electrical components in a vacant space at the Kariya

---

v. A prime example of this is the North American automotive and home entertainment industries which produced products of relatively poor performance and allowed Japanese imports to break into the home market.

vi. There is no agreement that manufacturing is going through a renaissance in the Western World. This will be discussed in detail later in the book.

plant.[19] With new imperatives on minimum use of materials, energy, and human resources, the business grew into what is now known as Japan's "largest manufacturer of electronic components for vehicles" - Nippondenso. It is one of the most efficient manufacturing businesses in the world.[20]

The aftermath of the war proved even more trying for Japanese manufacturers. World War II left an almost complete exhaustion of economic resources, and major plants and production facilities in ruin. In the spring of 1950, Toyota sent Eiji Toyoda to the United States for a tour of Ford's Rouge Plant in Detroit, hoping to find the secrets behind Ford's amazing production of 7000 vehicles a day. Upon returning and teaming up with Taiichi Ohno, both realized that the Ford system of mass production was too resource-intensive and demanding in capital to work in Japan, and they were forced to look at new ideas.[21]

It was also in 1950 that the Union of Japanese Scientists and Engineers invited W. Edwards Deming to Japan to help them through the postwar era. Deming introduced his ideas for a new system of management including the use of statistical analysis, the implementation of quality control, and the importance of the customer.[22] The state of Japan after the war, combined with the teachings of W. Edwards Deming and the vision of Taiichi Ohno and Eiji Toyoda, led to the evolution of mass production into the system now called Lean Manufacturing. This did not happen overnight. The first attempts yielded low-cost, low-quality products. More than 20 years of methodical improvements were required to make Japan one of the world's best producers of customer value, combining quality with affordability.

Lean Manufacturing

After Toyota mastered new methods of manufacturing, many other Japanese businesses adopted the system. The principle of the system can be captured in few words and it has to do with harmony and an Eastern vision of flow:

> Flow the right things, flow them quickly.

By the 1970s Japan became known for top-quality products at a competitive price and was rapidly becoming a leader among industrial economies.[23] The quality and reliability of imported goods forced North American and European plants into painful retreats from many profitable markets. Some industries never recovered; others started to adopt the principles of lean manufacturing in the early 1980s. This process is still taking place in many North American manufacturing industries: some are making rapid progress, and some are too late.

The idea of lean manufacturing may appear simple, but simplicity is a concept the human mind sometimes has trouble capturing. It took many years for lean manufacturing to become practical in North American society. The key to implementation was the silicon chip, which opened the road to a disciplined approach to manufacturing through automated machines and modern information systems. Silicon chips do not allow for the human desire to change, the human desire to improvise, and the human tendency toward disorder - they are intrinsically simple.

Automation and Robotics

Computers made it possible to think of manufacturing in a new light. Suddenly people were performing skilled and intelligent tasks while machines did the repetitive work with extreme accuracy. Accuracy is the reason it took 200 years after the start of the Industrial Revolution to see mass production without mass labor. When a product does not require accuracy, it is difficult to economically justify replacing men with machines. For example, today steel furnace "puddling" can be done by a machine, but only a smart machine that knows its way around the factory floor and realizes if it makes a mistake. Those machines are expensive. In most parts of the world it is still cheaper to have a gang of tough men puddling an open furnace than to buy expensive robots.

The robots came quietly, starting in about 1960. They progressed a little at a time, almost by continuous improvement. Most pre-programmed robots were originally used for repetitive tasks such as material handling, assembly, and inspection.[24] More recent robots can manipulate tools through complex feedback systems and adaptive controls, to perform a particular task, such as spot-welding or spray-painting.[25] The refined surgical-room robots, capable of neatly plucking the cornea out of an eye and attaching a new one in place, are a more genteel form of the same brutes that place large steel pieces together, but with microscopic accuracy. We will talk more about the significance of robots and automation later. For now it is important to state that the development of self-operating machines is the foundation of modern manufacturing. In Western society, it is this development that makes possible the adoption of new manufacturing methods.

Information Technology

The application of computers and robots in automation was driven by manufacturing engineers, highly knowledgeable in the technology, and skilled in portraying the advantage of substituting people with machines. This worked well in an environment where the prime directive was process efficiency, evidenced by the industries that benefited from this approach. A good example of successful business strategies based on an efficient automated process is the aluminum beverage can. The can has consistently been reduced in price from its introduction in 1970 to today, making it very difficult for competitive substitutes. The can manufacturer and their material suppliers can adopt this kind of strategy because of constant productivity improvements. In other words, they have created a competitive weapon in process excellence. Other businesses, such as some electronic hardware companies, are based on highly competitive assembly and distribution processes. These examples of outstanding competitiveness due to productivity are not the norm in manufacturing.

Most businesses can use manufacturing as a competitive weapon, but also need to add other dimensions such as agility in manufacturing and product design response to changing consumer preferences.[26] These other dimensions foster a different application of computers and networks. This is the use of information technology in the planning and managing of resources throughout an entire supply system. However, it took some time to achieve this.

Information technology (IT) in the 1980s followed the same route as industrial engineering did 20 years earlier. In the beginning it focused on the functional objective of gathering and ordering of information for its own sake.

Failures in early IT applications in resource planning and accounting systems aggravated the position of Western manufacturing. The reports these systems generated, disconnected from the needs of the user and from real time, became another competitive hindrance to Western manufacturers against the Japanese world of simple methods focused on the essentials.

Information technology is now more mature. Links with machine-generated data make the information more relevant in real time. Suppliers and customers can now do business in real time through networks. This has once again changed the vision of flow in manufacturing.

> Mass production envisaged flow of materials and products; industrial engineers added a time dimension to the application of labor and connected cost with machine time. Information technology and networks are bringing all these elements together in real time.

The Future?

Some scientists believe that the time is right for a new manufacturing revolution and that artificial intelligence is around the corner, meaning that one day humans will never again be involved in mass manufacturing. Higher intellectual processes such as framing information into a context, generalizing, and learning from past experience may soon no longer be the exclusive domain of humans, thanks to artificial intelligence.[27] Scientist Kevin Warwick believes the intelligent factory may be the way of the future.[28] Opponents to this view have raised questions about the limit of scientific discovery, and the disappointing applications of artificial intelligence thus far.

Our own experience suggests that the intelligent factory will not require new discoveries, only evolution and adaptation. The robots will evolve by small steps, and we may not even notice.

Where will this leave craft manufacturing? Today craft manufacturing in industrialized countries is not large in terms of revenues, but revenue only measures dollars of output, not value.

To find the real value we may want to reflect that it is craft production that creates the prototypes for the styles of tomorrow and generates most new technical inventions. Academic and industrial research is mostly craft-based and so are cultural innovations, which drive new designs (see Figure 1.1). Craft production may be the only form of manufacturing in which humans will be physically involved in the future. The rest will belong only to the machines.

Figure 1.1 - The Evolution of Manufacturing

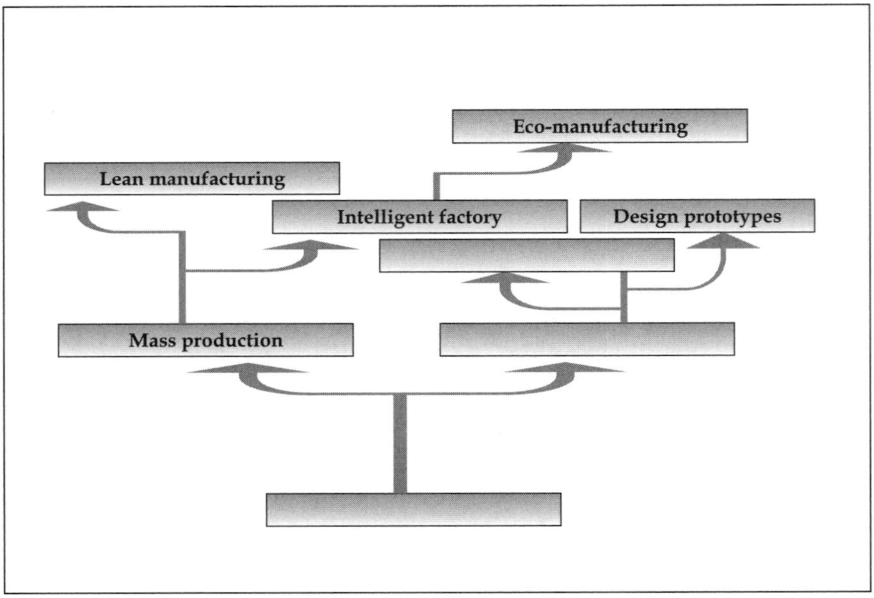

The need to look after the world in which we live is also something that could change the manufacturing processes of the future. Environmental awareness is becoming a large facet of today's society. Some, including Swedish scientist Reine Karlsson, believe that:

> in order to enable sustainable product development and to guide industry to adopt eco-efficiency and life-cycle thinking into commerce, there is a need for readily understandable system assessment tools.[29]

The manufacturing industry has tools at its disposal that it could use to improve the odds of sustainable development in the face of environmental challenges, although these tools might be considered mysterious and complex. The argument of Karlsson is for clarity and transparency, and for broad use of studies of alternatives. With or without the guidance of well-informed computers, the manufacturing industry is sure to change in some form in the future to meet the needs of society, as it has done for centuries. Some of the manufacturing principles we will discuss are an integral part of this change. They benefit the business and at the same time reduce the use of resources.

Although the history of manufacturing unfolds in a tidy chronology, it is impossible to describe the evolution of manufacturing from craft production to full automation in linear terms. In biology, the linear model has been supplanted by the "bush" idea where many systems exist together before reaching extinction, some of them finding new vigor in a different role. We have tried to represent the coexistence and complementary roles in Figure 1.1.

The branches may not be complete; they are meant to provide a structure for thinking about industrial evolution and for understanding the value associated with the coexistence of many approaches to the making of goods. Some of the upper branches are characterized by a higher content of information and of dynamic response. This makes them different (but not more important) than the branches based on individuality and creative design. We will explore these relationships in greater depth in a later chapter.

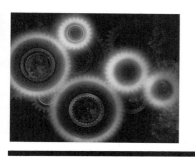

Chapter 2

Dynamics in the Marketplace

Introduction to Supply Chain Management

In the last few years of the twentieth century, the business process called "Supply Chain Management" (SCM) has gained prominence in almost all manufacturing and distribution businesses. There are now enough variants of SCM to fill an entire business school library, yet the principle is simply good business sense.

The business intent of SCM is based on the idea of strategic alliances, a concept introduced by Arthur Sloan in the 1930s. The technology of SCM is born out of the field of logistics and industrial engineering, but the incorporation of a customer focus and the inclusion of money and information with the "flow" of materials is outside the mathematical rigor of these disciplines. The business content and the expansion of the technology justify the shift in terminology from "logistics" to SCM, and perhaps to Integrated Supply Chain Management (ISCM).[i] Do other refinements in terminology reflect a fundamental change in understanding or even a real difference in content? Perhaps they do in some aspects of focus, but certainly not in the general concept. We will use the term ISCM in this book to underline the integrative nature of this business process.

Technically, ISCM is concerned with getting the raw materials to the firm's manufacturing plants, and getting the finished goods into the hands of the customer. In a single-step supply chain, these tasks are integrated into the function of the company.

Often, however, the manufacture of goods is not performed entirely within one firm. In a multistage supply chain, there is more than one firm involved in delivering to customers the product that they choose. This raises a number of issues with the flow of materials and information across company "borders," and it requires much more care with managing relationships. This becomes one of the key success factors in ISCM.

---

i. ISCM is a process-oriented, integrated approach to procuring, producing, and delivering products and services to customers. ISCM has a broad scope that includes subsuppliers, suppliers, internal operations, trade customers, retail customers, and end users. ISCM covers the management of material, information, and funds flow. (Note that we add the word "integrated" to underscore the objective of integrating the many functions into the total process.)[1]

Three factors have made ISCM a chosen strategy with many manufacturers:

- Growing importance of the consumer
- Global competition
- Development of information technology and communications

Growing Importance of the Customer

Whether due to an almost limitless choice of products with similar functionality, or some other sociological phenomenon, the customer is enjoying more importance at the beginning of the twenty-first century than ever before.

The increasing importance of the consumer may also have a technological component: the advances in information technology (IT) and the World Wide Web. On the one hand, the consumer is able to acquire an education on the best choices that satisfy his or her needs using a simple home computer. On the other hand, the process that the customer uses to search the Web generates valuable databases of demographic preferences, shifts of interest, importance of certain goods versus others, and similar valuable marketing information. This information gathering about customer "likes and dislikes" (which is now possible through data mining) was before only available to manufacturing businesses for very high-value customers (usually large firms themselves). One could say that there has been a certain level of democratization of customers, which has come about thanks to the advancement of IT.

A response to consumer preferences is customization. Businesses that have succeeded in customizing without adding cost to the product over and above what the consumer is willing to pay are thriving. Customization is a dynamic process, requiring continuous customer awareness and redesign, since preferences may shift in a period of months or even days. Is the Web detecting preference shifts or causing them? We will never know since conventional causality principles do not apply to consumer choices.

A successful example of customization is a small lighting systems company, Valley Lighting, Inc., of Baltimore. This company has been successful in keeping ahead of intense competition by providing customer value with "the lowest cost solution and service." If this does not sound inspiring, consider the following. Usually the customer of a small lighting business perceives lighting as an accessory to a building and not as part of pre-engineering, as is the case with large architectural projects. When it comes the time to think of lighting the small project is short of money and the customer looks for budget cuts. Rather than providing 70% of the desired result for 70% of the initial budget, Valley Lighting prides itself in being able to provide 90% of the "feel" for 70% of the budget. The "feel" in lighting is a result of customization using standard, low-cost components. An airplane lounge requires a relaxed soft feel, a nightclub a visually stimulating one; an office needs functional but pleasing illumination. Customization in this case is the art of providing a perceived value of "feel" for the right cost. Valley Lighting, operating in a highly competitive market, does this by keeping in the forefront of design and estimating, using software that networks the designers and estimators with detailed component information. The same software converts an estimate into an order, delivery information, billing, and maintenance management.[2]

The "successful" cases of customization are successful in the sense that they not only create a new market or rejuvenate an existing one, but are also profitable. They have sparked the idea of "mass customization." Mass customization implies the application of dynamic consumer data directly to the manufacturing of the product. In this case the successful examples are few.

The growing importance of the consumer does not always imply greater power of the consumer: it simply creates awareness of market dynamics which was not previously available. A customized car may be a greater or lesser value from a functional viewpoint than a standard design. The customer will only know that his/her car is different. Perhaps this is enough.

Global Competition

The pressure for cost containment and the adoption of many initiatives such as lean manufacturing, just-in-time (JIT), re-engineering, downsizing, and right-sizing are a consequence of the intense level of competition seen in many industries. This competition has moved from local to global markets as the result of international trade practices, advances in transportation of goods and, once again, global communication networks.

For an example, let's consider a New York fashion accessories business, with a strong affiliation with local fashion designers. In two years the business has moved into competing with other manufacturers in Tokyo, Rome, Paris, and Hong Kong. The competitive activity takes place at international trade shows, virtual auctions, Web advertising, and electronic brokerages. The owner has no idea whether this is better or worse than tending to a local market. What she knows is that the local market is no longer there and she must either swim with international fish or become a tasty morsel for the competition.

When firms look for a point of differentiation and a sustainable competitive advantage, they now often turn to ISCM and consider what they can glean from best practices and what strategies they might employ within their supply chain to increase customer value.

Development of IT

The science of logistics is mathematically intensive, and the ready capability of spreadsheets, linear programming, modeling, and simulation have all been brought to the desktop. No longer is logistic science available to only the large firms that can support mainframe applications. Small enterprises can now simply determine economic order quantities, inventory levels, and the like using desktop spreadsheet software. They can also model the entire supply chain with relatively accessible technology.

Besides the mathematics, the connectivity that has been made available with the progression of IT and global networks has had a tremendous impact on the capabilities of supply chains. Being linked with suppliers and customers in real time offers the opportunity to market, manufacture, and deliver on a "unit of one" basis, fully exploiting the differentiation of the firm's products to deliver maximum value to the customer.

The Technical Nature of ISCM

Management of the supply chain is primarily concerned with the movement of the product along a product stream from a state of raw materials to finished goods, through whatever interbusiness boundaries might be in place. The product does not have to be a physical entity; it could be a transaction, a service, a license, or the collection of income tax for the federal government. The principles of ISCM are the same for all these "products" and require discipline and rigor in execution. The rigor in execution has only become possible with computer automation and this marks the difference between the early manual efforts in JIT and modern ISCM.

Supply chains vary significantly in complexity: for some products, the process may take place primarily in one or two businesses, for others (such as the automobile) it may involve a very complex web of companies.

As was stated earlier, the roots of ISCM lie in logistics, which consists of the actual movement of goods considering all of the various spatial and temporal constraints. In the traditional field of logistics, the main concern is the delivery of goods in a timely fashion. The science of logistics was developed in the military, where the entire battlefield infrastructure is mobile and support, food, medical equipment, medical personnel, fighting personnel, and weaponry all have to be delivered to the battlefield in a schedule that will promote victory.

The science of logistics is mathematically intensive because it attempts to describe the integrated effect of multiple variables. The human mind completes this process quickly by intuitive approximations. An example would be the decision to cross a busy road, or the decision to get married. These problems, which appear relatively simple, are next to impossible to solve mathematically.

The best known of these difficult problems is the "traveling salesman," which illustrates well the point we are making.[ii] In this example, the object is to optimize the route taken between the cities that a salesman has to visit. To calculate the minimum length of travel for a five-city route, each of the 120 (= 5!) variations must be compared. That's not too bad. When the number of cities is increased to 100 or more, the computing time required to solve with this method goes to levels of billions of years! The advent of desktop software, which zeros in on narrower ranges of "better choices," has made approximate but adequate solutions to problems like this available to a much broader audience. The result is the capacity to do analyses and planning that was before confined to mainframe systems.

The science of logistics has been expanded (and renamed ISCM) to include the entire chain of events (physical or virtual) that extends from raw material production and/or procurement to finished goods sold into the hands of the end user. Examples of extension outside the traditional area of logistics include the supplier/customer relationship, the purchasing agreement, and the transfer of information and funds. Thus each "link" in the supply chain can see long and short-range forecasts, as well as actual demand, in as close to "real time" as is practicable.

The Strategic Role of ISCM

The future of ISCM is expected to include a more strategic role, as the functions of product development, marketing, and customer service are influenced through the expansion of IT and communication capabilities within the supply chain. Even now, ISCM can influence some product developments and designs (a new product design or material substitution may decrease supply chain costs, or increase efficiencies). As the capabilities within ISCM grow, more information will flow further up and down the supply chain and will influence more of these strategic areas.

---

ii. For a complete examination of the traveling salesman problem and the solution methods used, see E. J. Lawler, J. K. Lenstra, A. H. G. Rinnoy Kan, and D. B. Shmoys, *The Traveling Salesman Problem: A Guided Tour of Combinatorial Optimization*. New York: John Wiley & Sons Ltd., 1985.

In a slightly longer time frame, ISCM will influence corporate strategy as the potential for true "demand pull" comes closer to realization. A good example of this is the evolution of automotive businesses from massive integrated structures into distributed manufacturing and service, with large numbers of smaller suppliers. The new ISCM automotive businesses have substantially different corporate strategies as their focus is no longer on manufacturing, but on the task of supplying a product. Manufacturing is only one of the required steps in this process.

This book is about manufacturing, but we can look at the emergence of new retailing businesses and their capabilities in managing inventories and cash flow as a core strategy. This gives us a glimpse of what the future holds for manufacturing. Manufacturing is still in transition in the adoption of ISCM, particularly in products that are distant from the end consumer.

Customer Value

The question "What does the customer value?" is the cornerstone of ISCM, because the supply chain is designed starting from the delivery, not from the source. Delivering what customers value at a price they are willing to pay is also the key to sustainability of any business, and ISCM is one mean of achieving this.

We have already stated that the "want" of the customer is no longer a static attribute, as it might have been 50 years ago. The dynamics of desire are fast moving and few products can afford the characterization of "base commodity." Think, for example, of bread, a basic staple of diet for those in Western societies. Businesses that do well in this area compete continuously with each other in supplying the consumer with different grains, gourmet options, ethnic varieties, and convenient packaging. Only a few years ago bread in North American cities was a standard product, and one had to search for specialty shops in ethnic communities to get anything different. Indeed, the power of marketing in our consumer society is now directed to determine, not merely react to, the desires and expectations of the consumer. Determination may imply studies of user preferences, influence through advertising, incentives, or all of the above. It is an active role in creating value, rather than searching for solutions after the facts are established.

Satisfying customers in a sustainable way means making a profit for the business, as well as providing value. The two goals are not incompatible, but are difficult to reconcile without a long-term view of the business. Business sustainability relies on a "value proposition," a statement of what the firm can offer the customer in product functionality as well as price and service, what differentiates their offering from that of the competition, and what factors are in place to maintain a given market position. The questions that the value proposition must answer are:

- Who is the customer?
- What is the product, from the consumer's perspective?
- What is the compelling reason to buy?
- What is the key benefit provided by the product?
- What are the key points of differentiation?
- What makes that differentiation possible?
- What will sustain that differentiation/market position?

In the case of many consumer goods, the differentiation is difficult to find. Differentiation is often more "perceived" than real. Without entering into a dissertation on marketing and market segmentation we need to explain how this perceived value relates to manufacturing capability.

Assigning a "perceived value" to a product requires difficult judgments. Even with the best market analysis tools the information might not be complete; societal and personal values will be difficult to quantify, and the dynamics of all these factors will make single-point choices very short-lived.

To illustrate how the logical process of positioning the value of a product works, we have shown in Figure 2.1 three different products in the same market, having a range of prices and perceived benefits. A real example would be a home touch-tone telephone, a programmable telephone with memory and on-line services, and a business phone center with hybrid capability (internal server/external lines). The three phones are all performing the task of voice communication, but the business center will be perceived as higher value than the home touch phone or the programmable phone. If all three products are priced accordingly to their perceived values they will lie close to the 45-degree line of average value for money. (Please note that the total consumer cost is not the price of the equipment but the total cost of installation, service calls, maintenance, part replacement, and finally the equipment.) Telephones that fall below the average value line will be poor value, and telephones that fall above will be excellent buys. This type of graph can be quantified using market surveys, multiattribute analysis, and statistics of consumer patterns and preferences.

Figure 2.1 - The Value Judgement

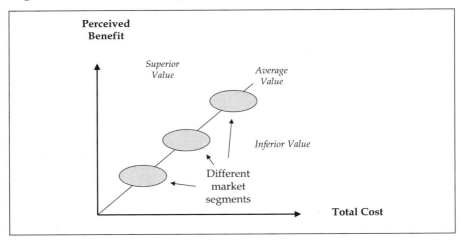

An example from the flower industry is Phal "White Dream" x "Morning Moon": a hybrid orchid of high quality with large, well-shaped white flowers. Orchids of this kind are produced through bioengineering techniques in specialized laboratories, and bear only distant relationships with their jungle ancestors. This orchid can be bought from a specialized orchid house for approximately $40 or from a general gardening center for approximately $12. The perceived difference between the two is that the orchid house selects the better specimens in the cross and the general gardening center will sell any surviving plant. Either of the two sources may provide a single plant with outstanding quality, capable of winning an award from the American Orchid Society or other regulating body, but chances are better with the orchid house selection. An award winning plant would sell for $500-2000 and would be promptly sent back to the lab to be cloned. Its clone, made by the thousands and identical to the winning plant, will sell for approximately $30. This is a great example of different forms of the same product resulting in different value and price on the basis of unique perceived properties. In each case, there is value for money.

The low-cost plant is a good fit for the unspecialized consumer who simply desires a flower. The orchid-house plant is for the collector hoping to win an award. The winning plant is for the commercial grower, who will then sell thousands of plants to amateur orchid growers, unwilling to give room in their greenhouses to anything but the best commercially available quality. The graph in Figure 2.2 represents the different consumer evaluations of selected forms of the same product. Only organized plant registry and worldwide communication have made this possible.

Figure 2.2 - Value and Price of Phal "White Dream" x "Morning Moon"

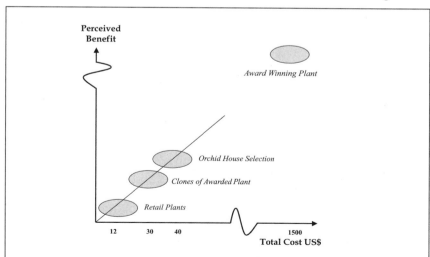

Forces in the Marketplace

If we assume that the product is priced in the correct market segment, and a functional supply chain has been designed, we still need to answer the question: "Will the new product be a success?"

There is no mathematical answer to the latter question (this is why we still need business managers with experience and intuition). Within each industry, there are specific forces that define the competitive battlefield. Michael Porter describes these forces as generalized principles in his book *Competitive Strategy*.[3] Figure 2.3 summarizes his theory.

Figure 2.3 - Michael Porter's Model of Competitive Forces

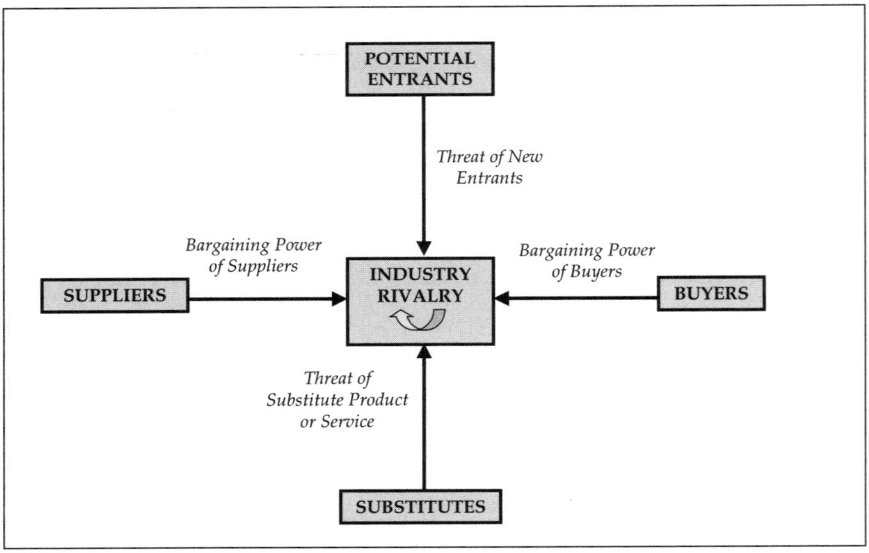

A detailed examination of the forces portrayed in this diagram will prompt a sound strategic plan, with good chance of success. Systematic strategic planning is a difficult exercise and businesses tend to shortcut this part of their activity, much to the detriment of results.[4] Dynamic simulation can do much to improve the mechanics of strategic planning, and to respond to Porter's concerns about analytical rigor.

At the center of Porter's five forces is the rivalry within the same industry, a subject of great focus in strategic planning. This may be the least important factor, as the behavior of the industry competitors is highly predictable, being in most part a mirror image of the planner's situation.

Moving to external forces, the supplier and the buyers have a great deal of power in some industries, and in others, none at all. For example, in the automotive industry in North America, one might guess that the power of the suppliers to the "Big Three" (General Motors, Ford, Chrysler) is limited. On the other hand, because cars are a consumer product, the power of the buyer is significant and leads to a constant pressure to increase customer satisfaction and value-for-money. Other examples are manufacturing industries making commodity metals such as copper, aluminum, and tin. In this case the power of the supplier is high, since the price of raw materials is often state-regulated in the countries where the ore is mined. The power of the buyer is also great because these metals are traded on international terminal markets and little can be done to influence a demand-driven price. These industries have only a few strategies for survival: cost control or downstream expansion into value-added products.

Another strategically important market force is the "threat of new entrants." If a new automobile manufacturer came into the marketplace in North America it would greatly influence the landscape of the industry, and could result in (further) questions regarding the survival of the "Big Three." The reason that there has been no such occurrence is that this industry has a high "barrier to entry," which is a contributing factor affecting the threat of new entrants. The capital investment in production facilities that the automakers have made in the latter half of the twentieth century and the beginning of the twenty-first century has helped to deter anyone from trying to enter the marketplace.

This may change in the future, if not in automobiles, certainly in retail businesses, because the Internet appears to reduce barriers to entry by promoting the growth of virtual businesses. It seems that all that is necessary to establish a viable business is to design a Web site. This simplified view has proven to be ineffective and Internet retailers are facing the fact that to be successful, they need more than a virtual site; they need an effective supply chain, and the ability to satisfy customers (back to our previous discussion on customer values, and the value proposition). This has proven to be the real barrier to entry into that marketplace. New technology companies are not exempt from this barrier formed by supply-chain effectiveness and customer service and satisfaction. Nor, indeed, are new products in an established industry. Take, for example, the developing reality of a home cogeneration fuel cell. The entry of such a product into the energy market will certainly be hampered by the scale economies realized by the current power generation and transmission companies, and by the level of service and reliability that is enjoyed by most of the Western world. Only if home cogeneration fuel cells can meet or exceed those levels of safety, reliability, price, and accessibility will they be considered a "threat" to the existing power suppliers. At this point it seems to be a significant, but not insurmountable, hurdle in the development process.

The last strategic force acting on a market is the "threat of substitute products or services." With each product offering, there is usually a competing technology or product or service that, although different in form from the incumbent(s), can satisfy the needs of the consumer. For instance, consider the airline industry. The internal industry rivalry is tremendous, and cost of operations is often the focus. But, while the customer has the option (usually) to opt for another airline, should there be a higher perceived value, he/she also has the option of adopting an altogether different form of transportation. By choosing train, bus, rental car, or boat, the customer has exited the industry to find a satisfactory replacement. There usually is a cost to do this, which is termed the "switching cost." Sometimes, the switching cost can be high enough that customers will be forced to accept what seems to be poor value-for-money. In the airline travel industry, for example, the switching cost may include punitive cancellation fees and the cost and inconvenience of extending travel time.

Porter's five forces are a comprehensive description of the issues to be addressed in planning a business. Business is not more complex than this; any other issues are supporting details that do not add conceptual value. Porter's model can provide remarkable insights because it forces consideration of a firm's role in an industry and the threats and opportunities that lie therein.

In addressing flow and dynamic effects, it is noteworthy that much of the information that is contained in Porter's framework is focused on changes in an industry and not on a static accounting picture. Many corporate annual reports now reflect this forward view and recognize the potential for changes in the business landscape.

Everyday Dynamics in Business Operations

Cases of drastic changes in the business landscape are strategically intriguing and make excellent studies for business schools. In the everyday operation of a business, there are other dynamic effects that greatly influence profitability and are sometimes equally as challenging to decipher as strategic trends. For instance, what are the effects of a two-year pricing or supply agreement? What is the impact on a business of a summer shutdown that leaves inventory short? What might be the effect of two stock-outs within a week for a retail outlet? What impact does a price reduction for the purpose of selling off stock have on the bottom-line performance of the company?

These questions point to one of the central ideas of this book. Much of the information that is reported as single-value variables to comply with accounting and general business performance metrics is, in reality, average values of statistical distributions.

The profitability of a business is increasingly dependent on the shape of these statistical distributions. This was not the case when communications were slow and operations were not directly connected with markets. Smart machines have changed this paradigm and created the need to interpret flow as it happens. Average results are no longer adequate.

How dynamic the performance of a business is depends on the type of product manufactured. A commodity may be highly volatile in price from day-to-day. High-tech products may be highly volatile in cost, because of variation in manufacturing.

The variation of price and its distribution is less intuitive than the variation of cost, but both are equally important. Cost is a statistical output variable of a manufacturing system that changes over time because each product manufactured uses a different range of resources, determined by the amount of materials, labor, and other inputs used to produce it.

For example, individual products that wait for a machine to be operational are more costly to produce than those that flow through directly, because some larger portion of the holding cost will be attributed to them. Likewise, those products that have more labor input due to a problem with setup or machine performance will be more costly to produce.

> Variation in cost is a consequence of variation in the manufacturing business process.

Price varies as well. The same product sold on a long-term-pricing or high-volume-discount agreement may be sold at a higher price to the short-term customer. Later we will discuss variation in price linked to automated machines. (An interesting example of dynamic pricing would be soft-drink dispensing machines with thermostatic sensors that automatically increase price with ambient temperature.)[5]

> Variation in price is a consequence of short- and long-term pricing strategies. Short-term strategies may be assisted by automation.

As price and cost vary, so too does the "bottom line." Profitability is more accurately reflected as a distribution or a range than as a single number. The location and size of that range will determine, to a large degree, the success and viability of the firm as shown in Figure 2.4. The maximum profit margin is the limit of the distribution, but this is only attainable in a few instances and is not an indicator of the potential of the firm.

Figure 2.4 - A Statistical View of Cost and Price

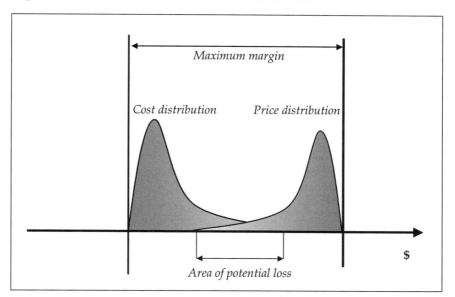

Variation along the Supply Chain

When a number of businesses are engaged along an integrated supply chain, the stream of activities result in the transformation of raw materials to an end product. While we can look at the value created in the supply chain within each of these firms in a sort of "front door to back door" framework, we can also take an integrated view of the chain.

Figure 2.5 illustrates the value created in the production of a forged aluminum control arm for an automobile. The value chain begins with the production of aluminum, and the casting of the alloy into forging stock billet. Those billets are cut and shipped to the forger, who forges, trims, and finishes the suspension parts and ships them to either the automaker or a "tier .5" supplier that produces the suspension system. The suspension system is assembled in the car by an auto manufacturer and the automobile is sold to the dealer and finally, to the customer. If one were to picture the value created for the whole chain as a pie representing gross profit from the sale to the customer minus the cost of the raw goods, the "pie" would have to be sliced five ways, as in Figure 2.5.

Figure 2.5 - Slicing the Value Pie along the Supply Chain

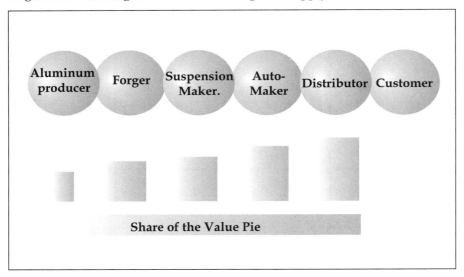

The way in which the pie is sliced is not a fixed mathematical formula. For the forged suspension it can be described as moving toward larger slices as the product gets closer to the customer, but this is not applicable in all cases. It is also important to note that the sizes of the slices change with time; in other words, there is dynamic variation in the division of revenues. Among the factors which drive this variation are delivery performance, sourcing changes, quality rewards/penalties, and discounts.

Dealing with Variation

The combination of all the described dynamic factors, variation in cost and price as well as variation along the supply chain, creates opportunities for continuous variation within the economic performance of a business. This should be reflected in the metrics that are used to describe or evaluate financial performance (some variation is not statistically significant and some is). Compliance accounting practice prefers fixed numbers, and the presentation of statistical distributions for this purpose would quickly become unwieldy.

It is the task of business managers to be aware of the underlying statistics in managing their business, and we will explore the methods accessible to them in an age of smart machines.

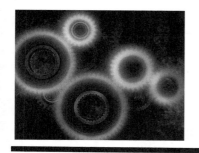

Chapter 3

Introduction to Product Streams

What Is a Product Stream?

Supply chain management does not guarantee that the manufacturing segment of a chain be lean and profitable. For this, a manufacturing operation requires specific techniques, which are all inherent in a concept we have termed "product streams":

> Product streams in discontinuous processes are a method of defining the manufacturing segment of supply-chain management and adopting the same rigor in passing attributes and information from one machine to the next as a supply chain would do with different business organizations.

Product streams transform the manufacturing segment of a supply chain in a way that simulates continuous flow. They integrate material flow with cost flow, capital utilization, and information flow in both productive and nonproductive steps of an operation. This is a more encompassing approach than traditional production methods. These methods concentrate on productivity and yield, both aspects of mass flow, but do not continuously measure either operating cost or capital utilization.

The concept of product streams originated in continuous chemical plants, such as refineries, where the physical aspects of the operation lead to ready visualization of product flow and cost. Product streams have always been applicable to continuous production plants because the process is designed into the plant's physical configuration and cannot change outside its control limits. These limits are set and monitored by automatic control systems.

Product streams can be designed for plants with discontinuous process steps if the previous definition is adopted, albeit with considerable challenge. Designing product streams in discontinuous plants was not practical until automation and integrated networks made the flow of information during manufacturing possible in real time. Real time refers to the present moment. Seeking real-time information on the behavior of a certain machine means knowing what is happening with that machine right now -

not what should happen, or what usually happens, but what is happening.

Real-time information provides a skilled worker with a picture of what is happening with particular machines at any given time in an easily understandable visual display. These displays show the present state of the process and the deviation from target, thus providing a continuous picture of process variation. There is significant difference between these real-time measures and statistical process control (SPC) charts, which are historical plots of a selected variable, such as productivity or yield. An SPC chart is useful in identifying long-term trends over hours or weeks; it does not show the instantaneous deviations from targets to which automated equipment can respond.

In addition, automation makes discontinuous processes more stable and allows a steady flow of resources because there is less reliance on human activity.

Product streams regard all aspects of the business in a statistical way while the operation is taking place by examining what is happening at that very moment. This makes it possible to minimize waste and cost at the time when a deviation from target happens. We also underline the importance of process stability. Correction should be exceptional interventions and not a continuous activity.

Product cost is a result of process activities in the product stream and, because of variation, is not controllable in historical accounting terms. Looking at last month's performance report is not a dependable way to predict this month's product cost accurately. Analysis of product stream data will clearly show that cost is a consequence of the stream operation and not an independent factor. We will discuss the results of recent research on the subject later.

Some manufacturers who have used detailed product stream management to reengineer their processes now produce goods with high customer value while simultaneously reducing all costs (including labor) to the bare essentials. The global economy will, in time, ensure adaptation of the same methods for other businesses to remain competitive.

The quantifiable elements of a manufacturing product stream, the "hard measures," are:

- Mass flow
- Cash flow
- Asset cost flow
- Information flow

There are several other elements of the stream that are not measurable in quantitative terms, such as skills and cooperation within and outside of the organization. We will deal with these "soft measures" in a later chapter.

Mass Flow

Active Centers

The concept of mass flow through a manufacturing process stream is similar to that of a chemical batch process. Material passes through a number of active processes where a transformation takes place (machining, assembly steps, logical treatment). We have called these active centers. Nonactive centers, such as warehousing, material movements, administrative practices, and quality checks, are interspersed throughout the active centers. We have called these nonactive centers, meaning that there is no physical transformation to the product, although they do sometimes contribute to the end value (Figure 3.1).

Unlike a chemical process, manufacturers face multiple options for process steps and sequences needed to achieve the same end. In other words, there may be several alternative process streams possible to make the same product, or multiple products may use parts of the same stream. This can pose a problem. A possible solution is to make the choices totally automatic, so that product streams are preselected and not driven by expediency. Numerically controlled machining shops are a good example of this approach.

Nonactive Centers

In several industries, including those producing automotive parts and semifabricated products such as sheet-metal, paper, polymers, and composites, automation and process control have increased mass flow through active centers. Stamping presses, rolling mills, and molding machines have become so efficient in productivity that "product-stream cycle time"[i] is largely controlled by the nonactive centers, which are rarely automated and exist as the major source of stream inefficiency and variation. The consequence is that productivity of major machines is improved, often at great cost, but the process stream does not generate any more output. This is a powerful argument in support of the measurement of mass flow of the total stream, rather than concentrating on local improvements focused on capital-intensive machines. Expensive machines also demand expensive upgrades and the end result is increased capitalization: a major profitability issue in modern manufacturing.

Figure 3.1 - Mass Flow in a Product Stream

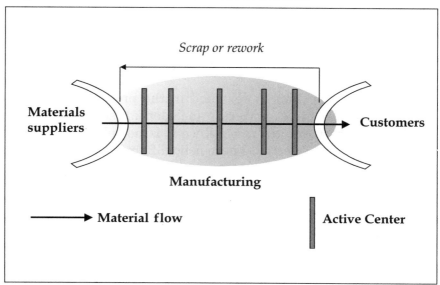

i. "Cycle time" measures the time spent by the product in one operation. "Product stream cycle time" defines the total time that the product spent in the system in both active and nonactive centers.

Generating Scrap

The reprocessing of scrap (known in automotive stamping plants as "offal") is often designed into the product stream; this is referred to as "engineered" scrap. In the automobile industry, scrap generated in the blanking of body components can account for 40-50% of the starter steel coil. Steel is a low-cost material and therefore scrap is of low value and not too significant to the product cost (the name "offal" is in itself suggestive of this). If the product stream of a stamping plant were extended to the steel mill it would be apparent that the cost of designing scrap into the process is great, since the capital-intensive steel mill may in fact produce only 30-35% of the material that ends up in the final product, the car. The rest either is internal recycling in the steel plant or is engineered scrap from the stamping press. Steel mills like to reprocess scrap, since it is low in cost and assists in balancing materials in the mill: an excellent argument provided that the scrap is externally acquired and not self-generated. The integrated cost of the underutilization of a steel mill asset is not always correctly reflected in product cost calculations, which may help to explain the marginal profitability of automotive steel producers.

Automotive companies are becoming aware of integrated product streams and are measuring the economic impact of capital utilization with methods like Economic Value Added (EVA).[ii]

In addition to scrap that is designed in the process there is also scrap due to defective material (accidental scrap). Defective material receives more attention since it is not planned, whereas engineered scrap is expected! There is no difference in the economic impact of either type. Scrap of any sort, whether planned or accidental, reduces the effective mass flow of a product stream in linear proportion with the defective or unused material. For plants that are capital or labor intensive, this has serious cost-absorption implications.

---

ii. This has stimulated the development of lower-scrap product streams such as nested blanking, tailor welded blanks, and other material-saving techniques. The use of life-cycle analysis would show the inefficiency of reprocessing 60-70% of all material. This technique is only occasionally used and is not yet part of recognized accounting practice.

Cycle Time

Once designed and understood correctly, mass flow (including reprocessing of scrap) can be controlled with the use of statistical techniques and advanced quality systems such as QS9000 and Six Sigma. These techniques encompass the whole product stream as well as all aspects of the supply chain. As mentioned before, the measures related to nonactive centers are less developed than productivity measures in active centers. Conceptually there is no difference between the two.

Plants that have started measuring product stream cycle time have found this a more intuitive metric for employees to understand in an integrated system. Cycle time is measured in hours used to produce a car, whereas mass flow is measured in cars produced per hour. Cycle time is easier to follow in nonactive centers, where mass flow is not physically visible.

Manufacturing plants that are accustomed to measure units produced per hour on their most complex machines would be well advised to also consider the total cycle time from the receipt of materials to customer delivery. This is a driver of product cost. Cycle times and all efficiency measures related to them must be based on calendar time. The science of industrial engineering has developed efficiency definitions based on planned time, or available time, which in some cases have the effect of reducing the calendar year to a little over 200 days. These definitions are meant for analysis of individual centers and not for management purposes. It is true that mass flow and cycle time efficiencies look a lot better when measured against planned time, but does the customer really care if a plant had a breakdown, or a planned maintenance, or did not schedule production in a certain week? Supply chains work in real time, 365 days a year.

> For management purposes, product streams must be measured as part of the supply chain, in calendar time.

Bottlenecks and the Theory of Constraints

In the early 1980s, Eliyahu Goldratt pioneered the idea of concentrating improvement only on the part of the process that constrains the product stream, be it an expensive machine or a simple clerical operation.[1] Goldratt's "Theory of Constraints" outlined in *The Goal* is quite logical. It is also a difficult task to implement since, in the real world, bottlenecks vary in time and space dependent on products and process performance. Only in an ideal system where a single product is processed through a unique product stream is the Theory of Constraints (TOC) literally applicable. The critics of TOC have been quick to point to this shortcoming and are missing the point. Goldratt does not suggest a mathematical solution, he proposes a new way of managing a business, and this new way has a major impact on the use of resources in manufacturing. Thinking in terms of TOC leads to engineering solutions that improve the total system in a cycle of continuous improvements, consistent with lean manufacturing and the best quality systems.[2]

The design and improvement of product streams is based on the design of constraints. To illustrate the point consider the case of an orderly group of 16 identical cars traveling on a four-lane highway. Every car goes 50 kilometers per hour and must maintain a spacing of more than 0.5 car lengths against each other. At the start all cars have a two-car spacing between them.

The road now narrows to two lanes and the cars still go 50 km per hour. The new spacing is easily calculated for two lanes: X = [12 spaces − 8 cars] / 7 spaces = 0.571. The cars are a little closer to each other but can still travel at 50 km per hour and respect the rule of >0.5 spacing. To make life difficult, the road now narrows to one lane: the equation above has a negative answer which means that two bodies have to occupy part of the same space. This is a bottleneck, which reduces mass flow through the whole system (Figure 3.2).

Figure 3.2 - Merging Destinies

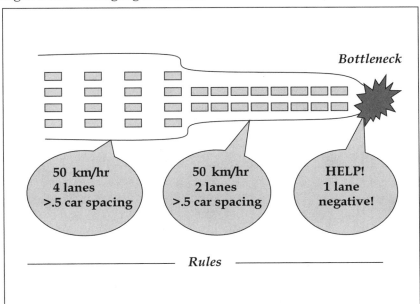

One lane is a bottleneck in the highway, only because of the constraints that are imposed on the traffic. These could be changed. For example:

1. Could allow a speed of 100 km per hour through the bottleneck.

2. Could impose a spacing of four cars length on the four lanes.

3. Could have posted a speed of 25 km per hour on the four lanes and 50 km per hour on the one lane.

4. Could have done all of the above and more.

Some of these solutions are more desirable than others and a good engineer would do a weight analysis, which would be similar (albeit more quantified) to that shown in Figure 3.3.

Figure 3.3 - Releasing Constraints

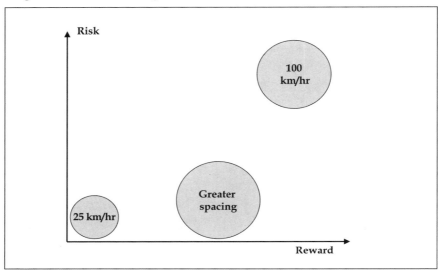

The solution of lower speed in all four lanes is low risk, but likely to be adopted by few drivers. The greater spacing is more acceptable, and the 100 km per hour solution is too high risk, even if most drivers would like it. This is a simplified version of what, in reality, would be a complex risk/reward analysis done by modeling and aided by optimization algorithms. The important lesson is that bottlenecks are not inherent in a system; they are created by constraints imposed on its operation, and constraints can be changed.

> This is the essence of TOC: do not reengineer the process stream or build a new plant, remove the constraints first.

Cash Flow

The economic contribution of the individual active and nonactive centers defines the cash flow of the operation.

A few preliminary definitions are needed to understand how money flows through a process stream:

- The measurement of cost must be in real time (in other words, it must flow with the product).

- Variable costs, which are directly attributed to the product, such as materials and supplies) are easy to measure in each center. For example, in making a car fender the cost of steel per fender is clearly attributed to the product at the beginning of the process, and the cost of shipping 50 fenders to the customer is clearly attributable to the product at the end.

- All other costs are more difficult to define. For example, labor cost is only variable in proportion to the shifts worked, but not to the number of parts made. Eliyahu Goldratt suggests that assuming all costs are fixed with the exception of materials and supplies makes the best business decisions.[3] Whether we accept this view or some other definition, the portion of cost that does not vary with mass flow must be allocated to the product. The most convincing formulae for this are based on the length of the production cycle for each product flowing through the system. Activity-based accounting is one of these formulae and has the advantage of being consistent with generally accepted accounting practices (GAAPs). Reconciliation between product streams and compliance accounting is thus respected.

- During manufacturing, the product accumulates cost. For this reason, manufacturing plants are best viewed and managed as cost centers, and not as profit centers. It is only at the very end of the stream, when the product is sold and the money received, that a positive economic value is generated (ideally higher, on the average, than the cost of producing the product) (Figure 3.4).

- The total cycle time from receiving materials to payment for goods shipped is the stream cycle time and is an important performance metric of cost and cost variation.

Product stream value contribution is a dynamic measure and cannot be completely reconciled with compliance accounting, which is done on historical measures. One is not a substitute for the other. Stream contribution is a daily management tool and is good for product stream design, process simulation, and trend analysis. Recent work in this field has brought discrete event simulation and activity based costing together and created excellent tools for cost flow.[4]

Figure 3.4 - Cash Flow in a Product Stream

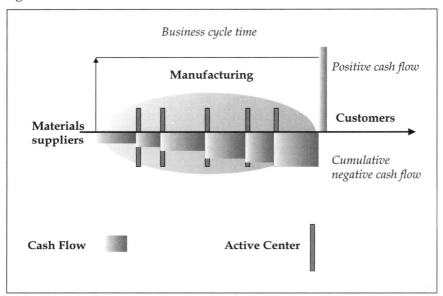

The direct measurement of cash flow is not a necessity in managing a manufacturing system, but is very useful in simulating the limits of system performance. As we will see in this chapter, the measure of cycle times and of mass flow bear a direct relationship with the cost accumulated by the product. Mass flow and cycle time are measured continuously in real time by automated machines and are easily understood by operating personnel.

Asset Cost

The value of the assets used in manufacturing contributes to product cost but is not usually measured at the plant level.

Appropriate valuation of the capital employed is done in both compliance and product cost accounting as a modification of plant cost. The result is that asset cost is not "felt" at the plant level unless the idea of an integrated product stream is implemented.

The capital used in manufacturing is divided into two categories:

- Fixed capital, which represents the physical assets employed
- Working capital, which represents the capital employed in inventories, bills payable, and invoices receivable

Depending on the business the capital employed can be very large (as in most manufacturing operations), or can be small (or negative) in distribution systems. The cost of the capital employed is a continuous function (banks never sleep!) and is particularly difficult to match with products going through discontinuous processes (Figure 3.5). It is, nevertheless, a critical element of cost accumulation and is poorly understood in many manufacturing plants. New performance measures, which put heavy emphasis on capital, such as Economic Value Added (EVA), have largely corrected this.[5]

Figure 3.5 - Asset Cost in a Product Stream

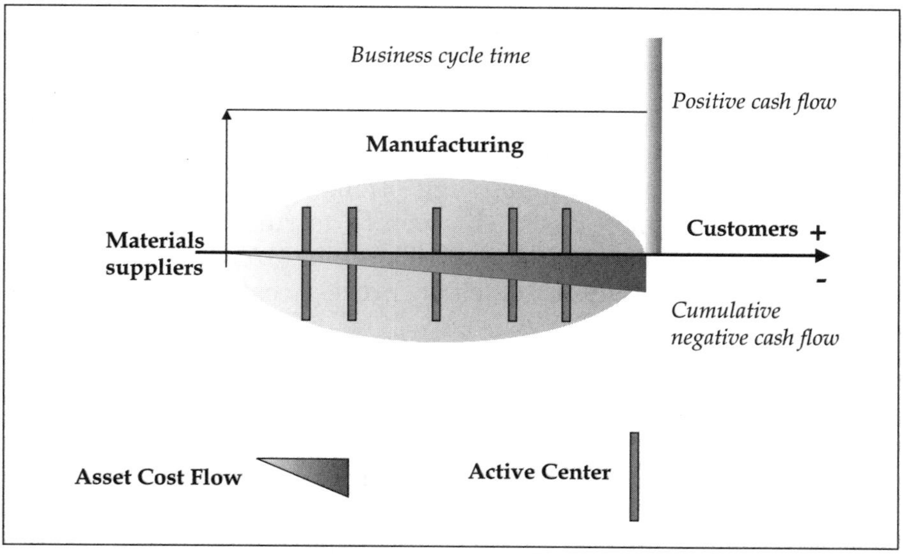

To show the importance of capital, consider the simple example of a person who owns or leases a $40,000 car. Assume that interest on $40,000 is $4000 per year. This is equivalent to $0.50 per hour, day and night. The driver is not likely to drive the car 24 hours a day, although this would be the best utilization of capital. Driving 2 hours per day puts the cost of capital at $6 per hour of car use, more than all the other costs of operation. Taxicab companies know this very well. They share the same car and drive three shifts a day.

The same idea applies to a plant: the cost of capital runs continuously. The less time a product spends in the plant (or the shorter the cycle time) the lower is the cost of capital that needs to be absorbed by each finished unit of production. Henry Ford realized this in 1913, when he changed his plants from two 9-hour shifts to three 8-hour shifts, maximizing the use of his capital and enabling him to raise the worker's wage to $5 per day while reducing product cost.[6]

Allocation of fixed and working capital to products is, at best, an accounting nightmare. It may also not be very significant, since the important thing from a management viewpoint is the total capital employed in the operation and its economic productivity. For this reason the approach of Goldratt, in assuming all costs as fixed with very few exceptions, has much to be recommended.

A practical production manager once saw a young industrial engineer struggling with capital cost allocations. To help, he put forward a refreshingly simple suggestion: "Product cost" he said, "is everything we have spent this month to run the plant divided by the units we have produced." This was 15 years before the introduction of TOC and the idea of "process costing."

Experiments in activity-based accounting, flow-through cost modeling, and other techniques each eventually reach an answer not that different from the practical engineer's simple math. The reason has to do with variation. Sophisticated allocation procedures are trying to measure something which is governed by the system variation. The only way that this can be achieved is to base all measurement on real-time information and not on standard hourly costs.

Information Flow

A discontinuous process will not automatically allow information to flow continuously in both the upstream and the downstream directions. The flow has to be specifically designed for each process, starting from commercially available planning tools. The design of the product stream flow involves building virtual operations for the nonactive parts of the cycle and creating a continuous chain (Figure 3.6).

Figure 3.6 - Information Flow in a Product Stream

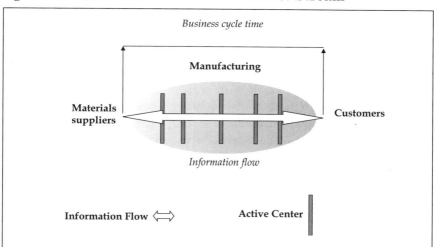

For example, a blanking press and a stamping press separated by inventory can be virtually connected by designing a traverse time for the inventory and using the inputs and outputs of the two machines to measure the overall cycle time and compounded yield. If the inventory traverse time is automated through automatically guided vehicles, it will ensure good repeatability. However, even with manual operation the virtual link can advise the operator when to affect material movements and improve overall performance.

Information technology in streams is best based on networks. Centralized data collection is inconsistent with flow design, and more difficult to manage than distributed information systems.

Flow, Yield, and Product Costing

The mathematical relationship between product cost and individual processes in a discontinuous manufacturing stream is fairly complex and is not readily definable without the use of statistics.

At first approximation, the cost of a product (Cp) is the sum of the cost of all materials used at each step (cm), the cost of all the additional resources used at each step $(cv$ or variable cost), all the allocations of the fixed cost in each step $(cf$ or allocation of fixed cost), plus the cost of inventory for the duration of each step $(cc$ or carrying cost). This is an approximation because:

- The cost of the net product must reflect the cost of reprocessing or reworking defective or scrapped material. It is impractical to accurately determine where the actual loss of production occurs and its economic value. If the loss is before a bottleneck or there is no bottleneck, the impact is less serious than if it is after a bottleneck. (An example is the trim on the edges of paper rolls, which have minimal impact on product cost, as they do not affect mass flow and can be easily recycled.) Yield also varies stochastically from one run to the other.

- Any costing method must provide for an allocation of fixed cost (a more accurate term is indirect cost, but it is simpler to think of costs as fixed and variable). This allocation is often based on individual machine times, with the resulting cost driver Cf per unit of machine time.[iii] There is no guarantee that this procedure will result in the full absorption of all the fixed costs, because cycle times vary statistically from one product to the next, even if the product is identical, and the rate is always calculated for a planned or forecast production level. Dynamic modeling, which can handle the stochastic aspects of cycle time, offers a more accurate estimate of how those costs are absorbed, and can also improve the accuracy of production level forecasts.

iii. The choice of drivers in activity-based accounting is discretionary and depends on the type of business. A business with very high material values may want to use material usage as the driver. In manufacturing the most common allocation is based on machine time.

Because the yield and cycle times are not definable as fixed numbers, activity-based costing cannot provide an accurate cost reconciliation, and the practice is to correct the balance at the end of a fixed period with an adjustment for manufacturing variances. This cost reconciliation must include all the costs of the business in a given period, where i is the number of the products and m is the number of process steps.

$$\sum_1^m Cp_i = \text{total business cost}$$

From the perspective of cost controlling, it is not necessary to measure Cp, which as we have seen is an output variable. The controlling input variables are shown in the following equation, where j is one of the individual processing centers in the stream, m_{ij} is the cost of material for product i in stage j, t_{ij} is the average cycle time of product i in stage j, cv_{ij} is the average variable cost per unit time of product i in center j, cf_{ij} is the average fixed cost per unit time of product i in center j, cc_{ij} is the average carrying cost per unit time of product i in center j, and y_{ij} is the yield of product i in processing center j.

$$Cp_i = \left[\sum_{j=1}^m (Cm_{ij}) + \sum_{j=1}^m (cv_{ij} * t_{ij}) + \sum_{j=1}^m (cf_{ij} * t_{ij}) + \sum_{j=1}^m (cc_{ij} * t_{ij}) \right] / \prod_{j=1}^m y_{ij}$$

This equation can be simplified in the following equation, where K_{ij} represents the sum of all the variable, fixed, and carrying costs per unit time of product i in stage j.

$$Cp_i = \left[\sum_{j=1}^m (Cm_{ij}) + \sum_{j=1}^m t_{ij} (cv_{ij} + cf_{ij} + cc_{ij}) \right] / \prod_{j=1}^m y_{ij}$$

$$= \left[\sum_{j=1}^m (Cm_{ij}) + \sum_{j=1}^m t_{ij} K_{ij}) \right] / \prod_{j=1}^m y_{ij}$$

This statement is confirmed by dynamic modeling of several manufacturing operations done at the Centre for Automotive Materials and Manufacturing. Cycle time, material usage, and yield drive product cost and can be measured statistically and controlled in real time.

In addition in most automated plants the machines generate the information, without any human intervention.

> In the case of a single product, going through the same product stream, the average product cost is a function of total material usage, average process cycle times, and yield.

Analyzing individual centers has little meaning from the perspective of business management, but is of great value for engineers when an improvement process is focused on a particular center. This is an important distinction, which is at the core of making a manufacturing operation a success. Managers must look at business systems; engineers must look at the details of every machine, with the perspective of the impact on the total system.

Product streams are an excellent way of achieving a broad vision for both purposes.

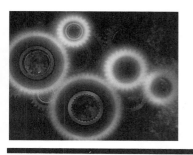

Chapter 4

Automation in the Factory

The Role of Automation in the Supply Chain of Manufactured Products

Automation in manufacturing plants has created a new paradigm in reliable production that is well beyond what is possible with human operators. The latest generation of automation differs from the machine control technology of even the 1980s in its ability to communicate through an entire supply chain without human intervention.

Humans show limited signs of having adapted to this new world of smart machines, particularly in manufacturing businesses where traditional management structures are leery of the new technology and unable to see the strategic opportunity that it offers. This is similar to the way many people could not see the impact of personal computers upon their introduction as business tools.[1]

Currently, the potential of the new systems is largely untapped by large-scale commodity producers. Smaller and more agile companies are leading the way into the next generation of manufacturing. With the advent of "intelligent" machines, there is already evidence that large manufacturing enterprises may fragment into smaller production units, linked as a virtual factory. This fragmentation has already begun in the production of automotive components where the role of the automotive company is evolving toward supply chain management and manufacturing is increasingly done by suppliers clustered in strategic locations.

While the evolution of automation continues, humans need to explore their new roles as problem solvers and designers of new methods, leaving the rest to automated processes. This is not an easy transition since it requires new skills and understanding of processes, machines, and business systems, accompanied by prompt decision making on the factory floor. The role of quality and control systems and the ability to adapt them to decision making in real time will be examined later.[i]

i. Quality systems, as applied in many manufacturing businesses, parallel in historical terms what the machine does already in real time: this suggests that the motivation for using tools like process control charts is one of compliance with customer and management requirements, rather than a real aid to problem solving and improvement.

The principles of product flow through supply chains have been well understood for at least 80 years. The motivation to build a new generation of "intelligent machines" is to embody these principles in machine design.

Examples are already apparent in the manufacture of many electronic products. In cell phone assembly lines, for example, machines can reach a level of accuracy, reliability, and communication with other centers that is unattainable with human operators.

There are challenges in creating a chain of manufacturing machines that communicate with each other and with the people conducting the business. Design and operation of these systems requires the bridging of different jobs, including control engineers, IT managers, and operations leaders. Most functional organizations can do this in a project mode (for example, during the design and construction of a plant), but are incapable of creating the required permanent relations for operation of the system.

In addition, the use of "intelligent machines" cannot be justified in terms of cost reduction or productivity improvements when the product has little technological content or is not closely linked to consumers. Many commodity businesses such as those producing wood products, concrete, and metals are unlikely to find the capital and skills required to retrofit new manufacturing technology into existing operations. In time, even commodity businesses may create different processes and organizations capable of handling automated supply chains, as many steel and aluminum businesses are now doing with mini-mills.

> There is a new generation of "intelligent machines," capable of connecting entire business systems. Their usefulness is predicated on the management structure and on the technological sophistication of the businesses.

Talking to the Machines

A generic model of automation in a network environment is similar to the structure of an individual robot (for example, a spot-welding machine on a car assembly line). The difference is the physical scale, which can encompass several factories, offices, and distribution centers, often separated by miles. A robot, as a self-contained unit communicating with a human interface, is easier to understand and construct. A chain of machines and clerical operations with multiple communication modes is a virtual system not perceptible as a physical entity and hence challenging to visualize.

The functionality of a virtual factory or enterprise can be pictured as a number of information and communication levels. Figure 4.1 is a typical schematic of a concept originally proposed by the Allen-Bradley company.

Figure 4.1 - Functional Diagram of a Virtual Factory

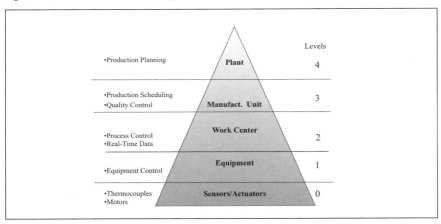

- The bottom level (0) contains machine sensors, activators, and preprogrammed sequences. It is strictly a local system accessible to technicians for maintenance purposes, but otherwise not visible during operation. (Some of these systems operate on a scale of microseconds.)

- Level 1 brings together signals from the sensors and

activators in a machine-specific control logic.

- Level 2 is still machine specific and includes the interpretation of production schedules received through a local area network, the processing of data from the control system, and the status of machine components. At this level, humans interact most directly with the machines, accessing operating information in real time as well as short-term trends, process control charts, and diagnostic procedures.

- Level 3 is a supervisory computer overseeing several machines and routing information to and from Level 2, as well as summarizing reports on performance. This level differs from the other two in that only a small amount of the information is in real time. The rest relates to discrete events, and to process or quality practices.

- Level 4 deals with the logic of the supply chain rather than the machines. Resource planning and utilization (Enterprise Resource Planning), order entry, queuing, and scheduling are done at this level, usually in an office rather than on the factory floor.

One could envisage other levels of more and more condensed and rarefied information, all the way up to the desk of the CEO. Some systems have as many as 14 levels.

> Although it is possible to build almost endless chains of computers and information levels, designers should avoid the temptation of this increasing complexity.

The consequence of multilevel systems was clear in a case reported in 2001, where the city of Toronto was unable to collect $700,000 in dog-tag fees. The information system linking the machines stamping the tags to the fee collection office was so complex that only one programmer had the knowledge to operate it. This person was accidentally laid off, with the consequence that many happy dogs walk around with free tags in Toronto.[2]

The Self-Taught Machine

The more complex examples of automation in manufacturing are found in industries that combine "transformation processes" with material handling. Examples of these are integrated food processing and packaging, the creation of shapes from molten metal, and the manufacture of paper from pulp. This type of manufacturing differs from the assembly or disassembly process where premanufactured parts are put together or taken apart because a transformation process involves controlled chemical and physical changes in parallel with material movement. Whereas the material movements are easily quantifiable in precise terms, this is not the case with chemical or physical changes.

An automated line producing automotive air-conditioning magnetic clutches is a typical assembly process.[ii] Robots build the complex clutches by locating individual components, placing them in position, and then welding the required connections while the assembly moves on a conveyer system. The conveyer is totally enclosed in a controlled environment and there is no human contact with the product. Similar processes are used to make electronic and communications equipment.

The unloading of cargo from a container by the use of robots and the electronic sorting of post from a mail sack are disassembly processes. Like assembly processes, disassembly processes do not involve a change in the state of the materials. Returning to transformation processes, an aluminum mini-mill (which produces flat sheet from scrap metal) provides a glimpse of the complexity involved. The process has several steps, all carried out in the same factory. First, the scrap is melted in induction furnaces, then mixed with pure aluminum metal and alloying elements in holding furnaces.

---

ii. A magnetic clutch is used to connect the air conditioning compressor to the engine drive, when the driver turns the temperature knob to AC. It is not a component of which drivers are particularly aware, until it starts to emit loud squeaks at around 35,000 odometer miles.

The molten metal travels to a continuous strip caster followed by rolling mills. The strip is finally coiled into large rolls of up to 20 tonnes in weight. The diagram shown in Figure 4.2 indicates the changes that occur in the chemical and physical states of the material.

Figure 4.2 - State Changes and Material Movements in an Aluminum Mini-Mill

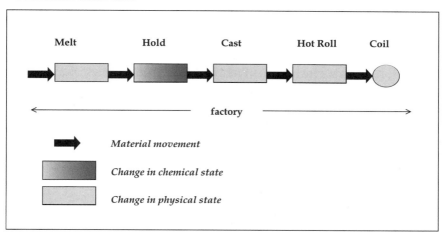

Aluminum mini-mills are highly automated and individual process areas are connected by fiber optics so machines can talk to each other. The structure of information flow is the same as that illustrated in Figure 4.1. Information flows continuously in both directions. This illustrates typical loops in control systems applied to an entire factory rather than to a machine. In the mini-mill:

- A machine downstream will be aware of what product is arriving next and will preset accordingly.

- A machine upstream will receive the results of processing the product in subsequent steps and also the deviation from required specifications and will make corrections to its setting the next time the same product is processed.

Through adaptive behavior, the whole plant will gradually learn how to produce better and better product attributes with minimum use of materials and energy.

This type of process design is common in the manufacturing industry. The few people involved in the operation assume a role of skillful supervision and maintenance of the machines and the communication system, rather than one involving physical effort. Product quality is no longer controlled by inspecting individual products, but by presetting the process and observing statistical variation of the machine parameters.

The design of the information loops is critical, not only to the performance of the operation, but also to the people performing different tasks in the organization. The information ladder in Figure 4.1 shows the logic of the automated mini-mill, starting from the machine-dominated sensor and control level up to the generation of reports and data for engineers, operation managers, marketing and sales, and finally customers.

The simple schematic does not portray the complexity of the system, which involves networks of hundreds of programmable controllers and minicomputers. Maintaining full functionality of a complex computer integrated operation is a special science. Continuous skilled attention is required to prevent the system from deteriorating. For now it is fair to assume the system will operate within the assigned limits of feed-forward, feedback, and adaptation.

Feed-Forward, Feedback, and Adaptation

Feed-forward is the process of presetting a machine for a specific task. It implies a detailed knowledge of the physical process in order to predict the relationships between process parameters and product properties. This is very complex in transformation processes. Feed-forward algorithms in metal processing are based on physical models incorporating mass and energy balances, state changes, chemical reactions, surface topography, and other phenomena specific to the operation. With the recently added capability of handling complex finite element analysis, these models are becoming more reliable at a steady state, but are far from making a perfect prediction during transient. They need the correction of several feedback loops in real time, as well as adaptation to repeated feedbacks.

Feedback is the process of measuring an output variable of a machine during operation, comparing it with a target, and correcting the machine settings to get closer to the target. Feedback involves a time delay between measurement and correction. It must be applied judiciously to avoid "hunting." It is therefore important for the feed-forward setting to be as accurate as possible. One of the best ways to design a control system is to make the machine adapt the preset algorithms to the feedback experience. This process is called adaptation.

The logic of feed-forward, feedback, and adaptation can best be explained with an analogy. Imagine a driver arriving at a road sign saying "Sharp bends for the next 2 miles." The driver, warned by the feed-forward sign, will apply brakes and reduce speed or clutch the steering wheel and hope for the best. Whichever option he/she chooses, the driver is responding to an expected change in conditions. An automated machine would do the same, only the information would be much more detailed. An engineer would have calculated the likely radius of the bend, the frictional conditions of the road, and the details in moment of inertia, suspension response, and wheel loading of the vehicle. The machine would then be able to predict an approximate speed to negotiate the bend safely.

Suppose now that the human driver enters the curve and finds that it is much sharper than expected. The driver may panic and go over the cliff. The driver may be skilled and use the accelerator to rebalance the car. The driver may simply skid through the curve while decelerating. All these are feedback reactions, similar to what the machine would experience in a much faster time frame. The machine would have sensors on wheel torque, which would automatically feed back into acceleration, attitude, and brakes, and would negotiate the curve with no trouble.

Continuing to drive on the road the driver is likely to read the next sign as a warning of "Very sharp bends for the next 2 miles," and drive with caution for a while. This is an adaptive behavior, which a machine would also adopt, by adding to its databank a range of sharp bends needing modified preset algorithms.

An automatically guided vehicle (AGV) negotiating difficult paths is a common feature in today's automated factories. Aside from flashing lights, soft beeping at intersections, and the ability to sing the company song, AGVs return to a feeding station when their batteries run low. The behavior of AGVs is so complex that people tend to regard them as intelligent pets and give them instructions rather than using their full capability. The earlier point about human interference with flow is appropriate in this case: AGVs could be preprogrammed to flow materials continuously as the product stream demands, and with no further instructions.

The role of automation in the management of product streams is one of reproducing a desired flow pattern through the productive and nonproductive cycles. This will only happen when the interface between human and machine has been resolved satisfactorily. In this respect, the factory floor in most manufacturing operations is no different from the frustrating experience of mastering the multiple-function buttons of an advanced washing machine or VCR.

> For most people the full potential of their washing machine will remain forever untapped.

The existing gap between humans and machines is better explained by looking at the development of automation over the years and the nature of self-taught machines.

From Cogs to Networks

While automation and material flow originated during the Industrial Revolution, the dream of the perfect self-operating machine, in humanoid form, dates back to the early Greek and Arab legends. In these legends, humanoids were crafted by men and given life by mystical powers. Details of the animation process have not survived, but the legends continue to capture the imagination of mankind.

Pygmalion, King of Cyprus and great sculptor, was known for his dislike of local women because of what he saw as wild sexual mores. Nevertheless, he fell in love with an ivory statue he had carved and called Galatea. Aphrodite granted him the ability to empower her with life. Pygmalion and Galatea married, and apparently lived happily ever after.[3] This is an unusual conclusion, since many other recorded cases of humanoid creations, including the ill-fated creature of Dr. Frankenstein, end in grief, disaster, and extermination.

Galatea's appeal may be waning in the twenty-first century because of competition from the realistic humanoid machines populating space dramas. Such machines do not require mystical justification to infuse them with life. They have been built by intelligent "scientists" and sustain themselves with everlasting "energy cells" or other "power converters."

Mankind's comfort with humanoids running around uncontrolled remains to be seen. In the Middle Ages, the chief rabbi of Prague built a set of humanoid figures out of clay, with the idea of protecting the Jewish community from pogroms. These figures, called golems, had superhuman strength and could be brought to life by inserting an engraved jewel in their forehead. When the golems rather predictably ran amuck, the rabbi was able to deactivate the threatening figures, after a long and courageous struggle, by pulling out the precious stones.[4] According to Kevin Warwick, Professor of Cybernetics at Redding University (UK), the golem story is a significant turn of events, because it introduces the idea of switching a machine on and off, a rather advanced concept at a time when the only power-driven devices were activated by water or wind and were not easily interrupted.[5] Controlled interruption is a fundamental concept in modern automation and the golems were not much different in principle from a modern robotic ore-mover, capable of superhuman strength and cunningly discriminating behavior.

The word "robot," meaning "worker," was first used in a 1923 Czech play by Karel Čapek entitled *Rossum's Universal Robots.*[6] Engineers have since adapted the word to depict an independent automated machine with some anthropomorphic features, such as articulated manipulators. More accurately, many manipulating devices used in manufacturing should be called "teleoperators," as they always involve human interfaces. Since it is difficult to imagine the word "teleoperators" becoming a part of popular vocabulary, with a small loss of accuracy it is likely that "robot" will remain the standard word for an automated welding machine.

Figure 4.3 - Basic Functional Blocks in a Teleoperator and a Robot

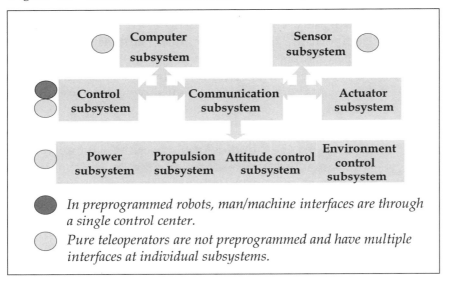

A detailed historical development of teleoperators is given in a paper by Edwin Johnsen and William Corliss, entitled "Teleoperators and Human Augmentation."[7] [iii] Johnsen and Corliss give a graphical description of the essential schematics of a "general purpose, dexterous, cybernetic teleoperator,"[8] which we have adapted in Figure 4.3, since a similar logic applies to robots.

This figure is a generic description of a complex and sophisticated automated machine capable of sensing and motion. Industrial automatons will not usually contain all of these features.

---

iii. Those interested in the more philosophical aspects of autonomous machines may want to read John Hugelend's *Artificial Intelligence.*

Analog and Digital Controllers

History's first computer scientist was Charles Babbage, a classic eccentric genius who in 1833, having already conceived a "difference engine," developed the first "analytical engine" which was a mechanical computer capable of accepting preprogramming in the form of perforated chips. Most important was the fact that the preprogramming was not limited to initial conditions, but could contain conditional branches, known as "if statements" in modern programming language. Babbage died without actually completing the engine, which operated on a system of gears stacked in vertical shafts and required extreme accuracy in construction. Imagine the tolerances and compensating mechanisms of a mechanical marine chronometer built into a machine the size of a modern diesel locomotive! Babbage started construction in his shop and the surviving parts suggest that, if completed, the engine would have been capable of operation and reliable performance.

Ada Augusta, Countess of Lovelace and gifted pupil of Charles Babbage, documented the principles and construction of the Analytical Engine, the first digital computer capable of any algebraic calculation. The Countess's writings suggest that numerals were expressed as 40-decimal digits (not binary) plus a sign, and the engine had a mill that performed calculations, a storage system (memory), and a sequence controller.

Calculators (some of them very elaborate) had already been invented in the seventeenth century and made additions and subtractions with ease. Some of these devices, such as Galileo's calculating compass, also contained built-in scales for radiant, degrees, and special-purpose calculations such as the size of powder charge in a cannon for a given projectile trajectory.[9]

In 1673, German philosopher Gottfried Wilhelm Leibniz added the capabilities of division and multiplication to calculators (in the form of the familiar slide ruler) by applying the newly discovered logarithms.[10] Although the slide ruler became an important device in engineering sciences and contributed to the development of mechanization in industry, building, and chemistry, it was not a computer. It was still a single-purpose device. The distinctive feature of Babbage's analytical engine was the fully programmable digital input containing any instructions.

The definition of a digital system is a "set of positive write/read techniques."[11]

- "Read" means recognizing an item by its position or its shape. The analytical engine did this through perforated chips - the keyboard of a computer reads when we press a specific letter.

- "Positive" means with no ambiguity. For example, a computer keyboard will always recognize the letter A as A.

- "Write" means manipulating an item by changing its formal position. Thus by pressing the letter A, a digital equivalent of the symbol is written in the memory of the computer and is displayed in the screen.

Many of our formal activities are digital, such as reading, writing, playing games, and building a house from a blueprint. We are immersed in a digital world in which it is becoming increasingly difficult to imagine a nondigital environment. One of the last bastions of the nondigital word, biology, is following the path of all other sciences by discovering the digitalism of genes,[12] the digital behavior of neurons, and the mechanical workings of all living organisms. The argument regarding digitalism is complex and will continue indefinitely.

Meanwhile engineers, with their digital minds, know very well what is digital and what is not and have adopted the word "analog" for devices that do not fit the read/write definition. There is even discussion of analog computers, a repugnant thought for mathematicians, but very real for engineers who sometimes need to simulate an approximate behavior without understanding the underlying causes, or need a simple, low-cost device to perform a single task.

A good example of an analog device is the mechanical thermostat based on a bimetallic strip, which changes in shape and causes an electrical contact to close or open at approximately the temperature set by turning a tensioning screw. There is nothing precise in this device, and it certainly does not read or write anything. Nevertheless, the expansion of the bimetallic strip, constrained by the approximate location of the tip of the screw, creates an environment for the operation of the device. A mechanical thermostat is useful because of its simplicity and low cost, but may now be obsolete.

Figure 4.4 shows the conceptual diagram of a mechanical thermostat and of a more recent digital device to perform the same function.

Figure 4.4 - Diagram of Analog and Digital Thermostats

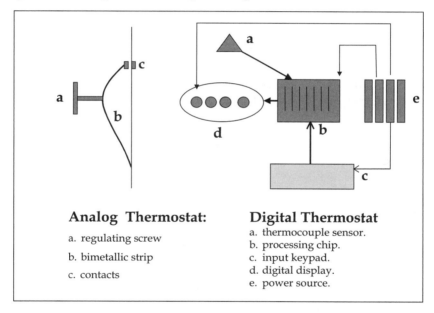

Analog Thermostat:

a. regulating screw

b. bimetallic strip

c. contacts

Digital Thermostat

a. thermocouple sensor.
b. processing chip.
c. input keypad.
d. digital display.
e. power source.

The digital thermostat is infinitely more complex than the analog equivalent and contains hundreds of parts in the four logical blocks. Its great virtues are:

- Precision, which can be tailored according to the application down to tiny fractions of a degree

- Reliability, which is implicit in the absence of moving parts

- Ability to communicate with other digital devices

The last attribute is so important that analog thermostats have probably seen their last days, as is evident by the mass production of digital devices.

There is still something elegant about the simplicity and transparency of operation of an analog thermostat, which a digital device cannot match.

Digital devices have difficulty in recognizing ambiguity. Lack of ambiguity is implicit in the definition of "positive reading." A digital device needs to read digits and anything nondigitalized is meaningless.

An analog device does not have this limitation and this may be the reason why there is an incentive to develop analog computers in either electronic or chemical forms. Organic computers in particular, using molecular mixtures capable of polarization, may be able to display more complex behavior than an electronic printed circuit, because they may provide not just "yes - no" solutions but all shades of gray in between.

The rest of this chapter will deal only with digital machines and their capability to approximate analog behavior by the use of mathematical techniques.

How Machines Learn

There is now satisfactory agreement that "intelligent" machines do not need to be built on an understanding of the human brain, no more than we would design a diesel tractor by understanding human mobility.[13] Machines can be taught or can learn by trial and error; the two methods relate to the concepts of feed-forward and feedback. This can be done in three ways: expert systems, neural networks optimization, and using genetic algorithms or similar logic.

Learning through Expert Systems

Expert systems provide a machine with a set of empirical rules to apply in certain circumstances. Expert systems add modifying rules to the numerical preset algorithms that apply when certain events take place. Those events are triggered either by an operator or by a signal from another part of the network.

In the mini-mill example, a machine in the system will receive feed-forward information from a previous machine, telling it that the next product to arrive is A and that to achieve the thickness required by product A, it should set loads, speeds, cambers, tensions, and cooling patterns to {2,17,11,5,35}.

Assume now that operators and engineers working with the machine have observed that better thickness control is achieved by slight modifications of the input set, dependent on product sequence, but are incapable of describing exactly how this happens. One option is to give the machine a set of sequencing rules and ask the machine to verify the results when a rule is applied (Figure 4.5). If the rule produces a better result, it will be retained. If the result is worse, the rule will be discarded.

This method ensures the systematic application of acquired knowledge in a way that no human can do. Humans will twiddle knobs one at a time and the process may be over by the time they have modified a setup. The machine can deal with an enormous number of variables instantaneously.

Figure 4.5 - Example of Rules for an Expert System

Product	Machine Set	Expert system
A	2,17,11,5,35	*If A start then change 2 to 3, else retain set.*
B	2,17,15,6,42	*If B after A then change 17 to 18, else retain set.*
C	3,11,17,8,50	*Always retain set*
A	2,17,11,5,35	*If A after C then change 11 to 14, else retain set.*

Expert systems are built in many automated machines and have improved reliability and quality. Their full potential has not yet been exploited for three reasons:

- In order to write the rules, experts have to agree—a major challenge!

- Some managers have interpreted the idea of "best practice" incorrectly as standardization of processes preset, thus eliminating the value of local experts in writing the rules and providing continuous improvement

- After the first creation of rules, maintenance and upgrades of expert systems is not done systematically, as the rules are no longer a high priority

Neural Networks

An alternative approach to expert systems, which circumvents some of the human frailties, is to design a system that will learn exclusively by trial and error and which can be "trained" to deal with a specific problem. Neural networks are a way of training and are extensively applied in automation to optimize setup conditions, either alone or in conjunction with expert systems.

The structure of a neural network is inspired by the functioning of a human brain, which contains approximately 10 billion neuron cells,[iv] highly interconnected through tiny filaments that touch, or almost touch, other cells. It is estimated that the human brain has approximately 10 trillion neural connections, operating in complex serial and parallel operations.[14]

Neural networks are constructed by simulating a single neuron and then connecting several neurons to form a network. A single neuron can be built electronically or mathematically. It consists of an input such as a series of pulses or numbers and emits an output, which has a nonlinear relationship to the input. In a mathematical neural network, the output can be a value emitted when the sum of the inputs reaches a certain level. Single neurons are connected in a structure like that in Figure 4.6.

iv. Some authors, among them Kevin Warwick, suggest the number to be closer to 100 billion. Both numbers are unimaginably large!

Figure 4.6 - Structure of a Neural Network

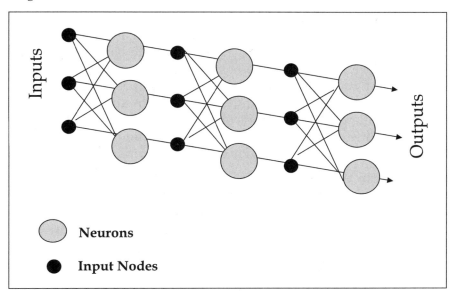

A system of this kind with intelligence equivalent to that of unicellular organisms can be trained to perform intricate tasks such as pattern recognition. Once trained, it will consistently perform the same way. The difficulty with neural networks is the task of training.

This explains why there is a practical limit to the number of connections. Training a neural network consists of reinforcing those connections that provide positive correlations, and weakening the relations that do not do so, in a fashion similar to that in which people train their dogs with positive and negative feedback.

The dog analogy stops here, since to "train" a neural network involves changing some correlation coefficient either with a manual control in an electronic black box or through a keyboard on a computer. By contrast, it is not necessary to open the dog's braincase to readjust its neurons.

Genetic Algorithms

Another way to optimize machine setup and improve continuously with experience is to collect all the information out of many runs of the same product and find which one gave the best overall result. The word "overall" means the best condition of many attributes, a concept that the human mind is able to comprehend only within two or three variables. This limitation does not apply to our bodies, which are shaped by thousands of evolutionary factors all acting at the same time. Genetic algorithms are mathematical approaches that mimic the genetic behavior of living organisms.

Introduced in the 1970s by Holland,[v] genetic algorithms (GAs) are now being widely applied to the optimization of processes and of supply chains. A GA is a model that evolves populations of solutions by means inspired from evolution and natural selection. Its logic is explained in Figure 4.7. The basic idea is to create an environment, governed by a set of rules, which kills solutions with a poor fit and reinforces solutions with good fit.

Figure 4.7 - Design of Genetic Algorithm

1. Find an initial population of random solutions
2. Evaluate each solution against rules
3. Produce k new solutions by: Selecting two parents with good fitness Applying crossover, mutation, or other genetic operators.
4. Add the new solutions to the population
5. Remove k solutions with lowest fitness and return to 3, unless stopping criterion is met

---

v. For further information on the origins of genetic algorithms, one should read *Adaptation in Neural and Artificial System*, by John Holland.

New solutions can be continuously bred by merging good parents and by random mutation.

In the GA terminology, solutions are encoded as chromosomes and evaluated according to a fitness function. The fitness evaluation is used so that the best chromosomes are selected more often for reproduction. Reproduction involves selecting two parent chromosomes and using a crossover operator to combine them into an offspring solution. Usually, a mutation operator is also applied to slightly modify the offspring solution.

None of this intricate work is visible to the user and interfaces have been designed to be very accessible. The key decisions are the rules governing the environment and the initial set of solutions, both of which are easily understandable.

A machine negotiating an unknown maze could be given an initial set of vectors and a set of rules. The vectors will give an initial "guess" to negotiate the maze; the rules will specify unsuitable solutions. A rule could be "kill all vector sequences which cause hitting a wall more than three times" or "keep all vector sequences which negotiate three legs in a row." By eliminating all unsuitable sequences and generating new ones, in the end the machine will negotiate the maze.

Results of the application of GA to either machine setup or logistics problems have been impressive. The technology is now making its way into many control systems. The only reason why the application comes so late is the amount of computation needed to zero in on a solution. Thirty years ago, it was an insurmountable problem. Today any laptop computer will run a complex GA optimization in a matter of seconds.

The Intelligent Machine and Real People

Intelligent machines will use a combination of the aforementioned techniques without the help of humans. In fact, our help is unwelcome. What, then, is the relationship between machine and people in an automated product stream?

One way to describe the relationship is to look at what machines do best and what humans do best and divide the roles accordingly.

Humans have unique capabilities in:

- Communicating with other humans (for example, customers, suppliers, and colleagues) and understanding their needs
- Designing products and building manufacturing systems
- Designing machines and their logic of control and operation
- Analyzing and correcting causes of deviation from the designed behavior

The machines are best at:

- Repeating the same task, with accuracy and consistency
- Optimizing performance within specified design parameters
- Collecting and formatting data on the product stream performance
- Collecting data and signaling deviation from the designed behavior

While the evolution of automation continues, humans need to discover their new role as problem-solvers and designers of new method, leaving the automated processes to do the rest. This is not an easy transition, since it requires new skills and an understanding of process, of machines, and of business systems accompanied by prompt decision making on the factory floor.

In an ideal world, the correct allocation of these unique capabilities will lead to excellent design and operation of product streams, no matter how complex. Exchanging roles between human and machines is not a good idea unless we wish to return to products of much lower value.

Meanwhile, in order to avoid the example of the Toronto dogs, it is advisable to beware of the three most common pitfalls of the virtual factory:

- Information technologists love processing data, not distilling information. If a machine generates immense databases processed through a resource management system, this information is invariably elevated to a level where it becomes confusing, rather than helpful. Because the language of the machines is not translatable into conceptual statements (it is simply a record of events), a machine cannot provide useful information to a manager without the interpretation of a user familiar with the process. Sending more data does not help.

- Designing the report that contains the conceptual statement is rarely done in consultation with the end user. Information specialists are usually asked to reproduce the format of conventional reports based on historical trends. They often miss the opportunity to portray real-time events, which is possible with intelligent machines. The fault does not lie with the IT specialists, but with the user's lack of understanding of the system capability. Users could see a much more up-to-date picture of variables which drive customer satisfaction and producers' costs in a supply chain. Virtual factories provide the opportunity to see manufacturing as it happens and the days of poring over historic reports could be over.

- The ultimate temptation is to help the machines to do their job. A good example is in batch processes, where products go through various machines and stop in intermediate warehouses. Given a production sequence, an automated factory is entirely capable of moving the products from one process to the other and even reacting to events such as machine breakdown, rearranging the sequence to optimum flow-through. The production manager should look only at the entry and exit points of the system and not attempt to intervene by changing a single machine queue or emptying a warehouse. If this looks like loss of power and control, think of how much more useful it would be to use the knowledge and experience of the manager to design better instructions for the computer instead of confusing the poor machine with irrational decisions. Networked machines learn and improve from experience, but machines cannot distinguish between an irrational intervention and a random event. They will learn bad habits in the hands of an unskilled master.

Avoiding these pitfalls is difficult in functional organizations; hence the need to carefully consider the structure of the people system when implementing product streams.

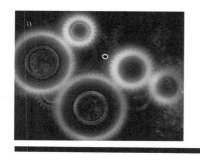

Chapter 5

Variation in Dynamic Systems

Variation in Static Systems

The human mind has a remarkable ability to aggregate large numbers and details into single concepts. The mechanism through which this happens is becoming more clearly understood and reinforces the profound differences between humans and machines, including the "intelligent" machines described in Chapter 4.

Humans see a pile of bricks and mortar as a house, a mass of different vegetable species as a forest, a box of eggs as a potential omelet, and so on. Rarely in our daily tasks are we forced to think of the details that lie behind the aggregated concepts.

When the need arises to build a machine or to "talk" to a machine, the detailed language of science is required and general concepts are of no use.

A machine is incapable of seeing a box of eggs as a potential omelet, and will not even see all the eggs in the box as being equal: some will be large, some will be small, some will be whiter than others. Humans, consciously or unconsciously, will eliminate a certain range of variation in egg color and size as irrelevant to the task of making a good omelet. To explain this task to a machine, we need to adopt a complex language that defines all the elements of a process in unequivocal terms, starting from detailed descriptions of the input materials and their range of properties, and then all the elements of each process step.

For example, if the eggs came from a local organic farmer, the variation in color and size would be greater than if they came from the supermarket where they had been put through a grading system. A machine may not recognize all eggs as the same input material unless we explain that a certain range of color and size is acceptable. In addition, if we wish to make an omelet, we will need to specify acceptable ranges for every process and material variable in the system. This is the logical process explained in product streams, and the language used to describe this process to machines is the statistical description of key factors governing the flow. These are called critical input variables (CIVs).

We can look, as an example, to the attribute of "whiteness" in eggs. In nature, the color camouflage of eggs is one of the survival strategies of wild fowl (a bird laying white eggs in dark sand would be very misguided). Domesticated poultry carry the genes of their wild ancestors and the practical result is an egg color distribution similar to those plotted in Figure 5.1.[i] There are two distributions in this figure: one for the farmer's eggs, and one for supermarket eggs.

Figure 5.1 - Distribution of Color in Eggs

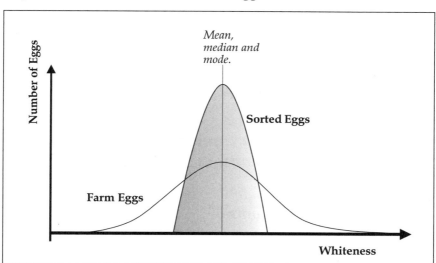

Looking at the two curves, we see two normal distributions of whiteness with the same average value. The difference is that the sorted eggs present a narrow range of whiteness, while the farm eggs cover a broader range of hues. In both cases, the average color is the same and, should we decide to use this value, it would not convey sufficient information to a machine. It is only the understanding of the variation of color portrayed by the two curves that will distinguish the two starting materials and allow us to

i. For those who want to know more about hens and the second law of thermodynamics, and the theoretical need for variability in color, there is now a reasonable range of accessible literature from credible scientists. See, for example, Paul Davies's *The Fifth Miracle: The Search for the Origin of Life.*

design instructions about acceptable ranges of color, leading to the perfect omelet.

The narrower distribution of the sorted eggs comes at the cost of processing them through a grading station (order does not increase without expending energy) and must be justified by the higher profit generated in selling boxes of beautifully uniform eggs.

The reality of an industrial product stream is much more complex than the static case we have discussed because it deals with multiple material attributes that change constantly during processing.

Understanding variation in a dynamic system such as a product stream is a challenge. A special science has been built around the statistical control of manufacturing processes and the resulting quality of manufactured products.

Manufacturing Industry and the Use of Statistical Process Control in Dynamic Systems

Variation and Quality Control

The stated goals of many manufacturing businesses reflect a desire to satisfy customers by "eliminating variation in production," or reducing variation to inconceivably low levels. We believe the managers of these businesses (and their professional advisors) are not suggesting a change in the laws of nature, nor are they willing to invest enormous energies to increase the order of things.

Our example of variation in a static population of eggs illustrates two important points:

- Complete order is not attainable (see second law of thermo-dynamics)
- Variation can be reduced, at a cost, with a specific objective in mind

The goals of a customer-oriented business are better restated with a more strategic purpose in mind:

> Variation should be controlled within limits which are acceptable to the chosen customers.

Far from being uninspired, this goal reflects a strategy that tailors quality to the requirements of a specific market sector and ensures adequate profits in manufacturing by containing cost. Should the business decide to pursue another product with higher quality requirements, variation would have to be adjusted accordingly.

When applied to a well-designed product stream, and with a clear vision of the target market, statistical process control produces excellent results. For example, the use of systems such as Six Sigma has had impressive results for product reliability and reduction of manufacturing waste in electronics, communications, and consumer durable products.[1]

Would Six Sigma techniques work miracles for an integrated steel mill? The answer is probably no, since the steel mill is only connected with a market through a long supply chain. Steel is not visible as a product to a purchaser of an automobile; hence, the scope for a clear quality goal with significant market impact is limited. In a steel mill (and in many other businesses dealing with semifabricated products), the use of statistical process control is still justifiable, but with a different goal, such as control of cost through reduced waste and rework in the whole supply chain.

Less desirable applications of statistical process control are those done for compliance reasons. Process control charts with points related to events of past weeks adorn the walls of many small and large plants that profess to practice quality systems. These plants are responding to customer requirements for qualification and have no specific targets for application. Training is done through a standard methodology, with little technical questioning of the suitability of the techniques for a specific process.

A good example is ISO 9000, one of the most used (and abused) quality standards in the world. Training in ISO 9000 and the installation of appropriate records is expensive, but ensures a banner celebrating the achievement for the front of the plant. This is where the process sometimes ends. We have seen qualified operations continue, undisturbed at the floor level, to produce in the same way as they have done for years, creating poor-quality products with wasteful processes.

Proper application of process control to a complete product stream is not common in manufacturing, and with good reason. Maintaining control is a grueling and never-ending race coupled with a continuous effort to ensure communication with machines. In businesses where customers are not visible as the prime motivators, few can sustain the effort for long, whether they are machine operators, maintenance technicians, or engineers. Think of the difference between a brewery where the flavor and appearance of the product are subject to frequent customer feedback, and a wood-pulp plant selling into a commodity market. It is easy to equate the two systems in terms of their process control needs, and it is not hard to argue the difference in motivational aspects.

The answer for businesses that do not produce consumer products may be in another generation of machines - machines that will take over statistical process control from humans. Some already exist and are probably a more economical answer in this case than the continuous training and motivation involved in the management and organization of people.

Quality systems and statistical process control were originally developed for consumer products and services. In these applications, they are effective in reshaping the attitude of an organization toward customers, resulting in improved market share and profitability.

For process-oriented industries, the same principles apply, but it is a better choice to build them into the machines. Until this happens, we are unlikely to see significant progress in quality and productivity.

In either case, the strength and limitation of statistical process control and other statistical techniques should be understood in order to ensure the best application.

Statistical Process Control

The concept of controlling a business and all the elements that create and sell a product is as old as the origin of manufacturing. Some of the earliest documents inscribed in stone do not talk of kings and gods, but of an inventory of goods and their value. Control implies knowledge of what is likely to happen next, and the generation of manufacturing in which we live is quite different in this respect from those of the past. When change in business was slow and fairly predictable (for example, in the supply of food produce), looking at past trends (including accounting data) was useful and allowed for an easy extrapolation of information.

Today, the rapid rate of change and fast diffusion of information makes the use of yearly and monthly reports a feeble tool with which to forecast the future. Donald J. Wheeler suggests that:

> managing the company by means of the monthly report is like trying to drive a car by watching the yellow line in the rear-view mirror.[2]

This strategy works fine while the road is a perfect straight line, but becomes difficult when the road starts to twist and turn.

If we cannot use accounting reports, then what can we use? The use of a statistical approach was suggested by Walter A. Shewhart in the 1920s, had limited application until the emergence of quality programs in the 1960s, and did not gather momentum until computers and networks allowed the examination of statistical data over a large span of business processes.[3] The important difference in manufacturing today is that the generation and analysis of this data can now be done in real time, and process control techniques must be adapted to this in order to be of real value. Looking at a control chart several days old is no different from managing a company by the monthly report.

Statistical Distributions and the Measurement of Variation

This section is a brief reminder of statistical principles, which are relevant to the control of processes. We will start with the random walk paradigm and the central limit theorem.

The Random Walk Paradigm

Most natural processes have a statistical distribution that is skewed in one direction. For example, the speed of a car on a police-patrolled road, if measured repeatedly, has a skewed frequency distribution because the car is not allowed to go higher than the speed limit, but has no limit on how slow it can go. Most drivers will opt for a speed close to the limit. The results of most natural processes are similarly driven in one direction from some limiting factor. This frequency distribution is explained in Figure 5.2 in an amusing analogy by Steven Jay Gould.[4]

Figure 5. 2 - Drunkard's Walk[ii]

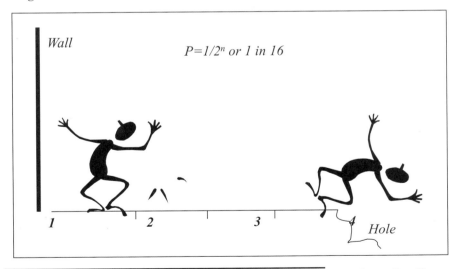

ii. Biological populations have skewed statistical distribution because of mortality. For an interesting description of the random walk paradigm applied to biology, see S. J. Gould's *Full House*.

A drunkard exits from a bar and decides to head home along a sidewalk that is 4 feet wide and bordered by a ditch and the wall of the bar. The drunkard weaves his way ahead, progressing one step forward and one step sideways either left or right. The wall is known in natural sciences as a "reflecting boundary" and the drunkard cannot help but make the next move in the opposite direction. The probability of each step ending in the ditch and terminating the walk is $1/2^n$, where n is the number of steps to the ditch. If we plot the positions of the drunkard over the random walk repeated a number of times, the highest frequency will be around the first step from the wall, which is the furthest from the ditch and still allows for two-directional movement.

In most manufacturing businesses, processes and transactions have reflective boundaries; some are caused by the measurement system, some by the nature of the process (Figure 5.3).

Figure 5. 3 - Skewed Statistical Distribution Caused by a Reflective Boundary

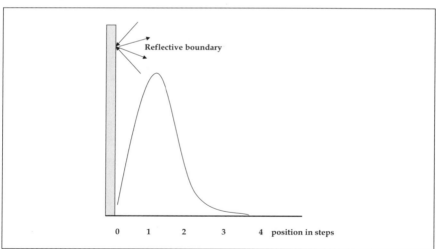

An example of a measurement boundary is the thickness control of paper in manufacturing. The isotope gauge used in these applications has a lower detection limit but no upper detection limit within the range of paper thickness, and this causes a skew in the collection of measurements, similar to the random walk.

An example of a process boundary is the declining performance of a bearing due to wear. Clearances can only move away from designed tolerances; they can never become better. The statistical aspects of measurements and process boundaries are quite similar and many performance indicators in manufacturing are of random-walk distribution. Figures 5.4 and 5.5 show cycle time distribution and variable cost distributions from a single product going through a sheet manufacturing plant with 36 machine centers and 22 product streams.[iii]

Figure 5.4 - Cycle Time Distributions for a Single Product through the Same Product Stream

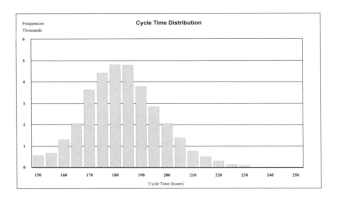

Figure 5.5 - Variable Cost Distribution for a Single Product Going through the Same Product Stream

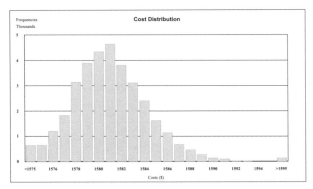

iii. This information is part of the Centre for Automotive Materials and Manufacturing, "Techno-Economics of Advanced Manufacturing Systems," Ontario Research and Development Challenge Fund project.

Skewed statistical distributions are not easy to handle in controlling a process, since neither the mean, the median, nor the mode (the point with the highest population of data) provides any information about the range of process performance. For this reason, control practice in industry is to convert data into a normal distribution, which is more amenable to analysis. This is the foundation of control charts.

Central Limit Theorem

The central limit theorem states that the distribution of sample means from a population of any underlying distribution will tend toward a normal distribution provided the sample size is greater than four. The larger and more random the sample, the more closely the means will tend to a normal distribution.[iv]

Process data converted to normal distribution will reveal whether the process is in or out of control in a dynamic way. If we look at the distribution of data from a brief period of process operation during two separate time periods, we will observe patterns similar to that in Figure 5.6 in the cases of a process under control, and a process out of control. In both cases, the process data are normal distributions and not single points, due to variation of the process and to the sampling method. This can only be changed by a physical modification of machines and measurement methods.

---

iv. With a sample of four random observations the confidence limit of the conversion to normal distribution is 1.5 sigma. This is occasionally referred to as a natural variation, but is in effect a measuring error.

Figure 5.6 - Process Under Statistical Control and Process Out of Control

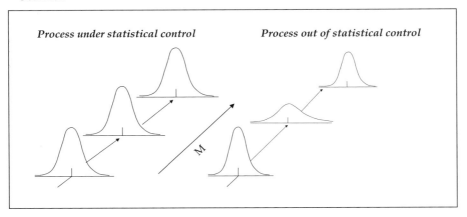

Process under statistical control   *Process out of statistical control*

The difference between the process in control and the one out of control is that the normal distributions vary with time instead of having the same shape.

If we want to express the shape of the curve as a measure of variability in time we can use two indicators:

- Mean: the weighted average of the distribution

- Standard deviation: an indication of the variance of a data point from the mean of the distribution.

The formulas for mean and for standard deviation are given in Figures 5.7a and 5.7b.

Figure 5.7a - Equation for Mean   Figure 5.7b - Equation for Standard Deviation in a Small Sample (<100)

$$\overline{X} = \frac{\sum\limits_{i=1}^{n} X_i}{n} \qquad\qquad S = \sqrt{\frac{\sum\limits_{i=1}^{n} (X_i - \overline{X})^2}{n-1}}$$

X_i = data point n = number of data points

$\overline{X}$ = mean of data points S = standard deviation

To illustrate this we return to our example of the box of eggs, and use a color meter to determine whiteness (100 is blue-white and 0 is brown). We will adhere to the following steps:

- Measure three batches of eggs

- Calculate the mean and the standard deviation in each sample

Figure 5.8 - Sample of Egg Data with Averages and Standard Deviation

	Sample 1	Sample 2	Sample 3
	55	46	59
	57	54	60
	54	56	68
	63	44	65
	56	60	63
Average	57	52	63
Standard deviation	3.54	6.78	3.67

The results are shown in Figure 5.8. Samples 1 and 2 have lower averages, but sample 2 has much greater variation. Sample 3 has a greater average, but similar variation to sample 1. From the point of view of the observer the eggs in sample 1 will be darker in color than those in sample 3, and sample 2 will look like a mix of many shades. If the observer wanted to make an omelet with moderately tinted eggs he would instinctively choose sample 1. The example demonstrates that both mean and standard deviation need to be measured to describe the statistical behavior of a system.

Shewhart's Control Charts

In 1924, Dr. Albert Shewhart developed a much-improved visual display of process statistics.[5] Working with Shewhart at General Electric, Edwards Deming realized the power of the charts developed by his colleague and encouraged the manufacturing industry to make use of them.[6] Deming's first great success came 25 years later in the Japanese manufacturing industry. These charts were not reintroduced into the United States to become the staple of every quality system until the 1970s. Today it is difficult to be in business as a manufacturing supplier without control charts.

A control chart is a time sequence of measurements, usually of sample means and of sample ranges.[v] The chart has a central line, which acts as a reference to detect shifts and trends, and two equidistant control lines called upper and lower control limits. These limits are mathematically defined from the sample population and are an expression of variation due to measurements and to common causes. A process that varies between the upper and lower control limits is in control regardless of whether its output conforms to customer requirements and whether or not it produces a manufacturing profit. It is important to understand this point since the appropriate corrective action depends on it.

By common causes, we mean that the design of the process and/or the management system is responsible for the variation. This includes the sampling method. In the case where the central limit theorem is applied to populations of four samples, the variation due to approximation is 1.5 sigma, as stated before. Besterfield[7] has determined that 80-85% of variation in quality is due to common causes such as manufacturing system design, and that less than 20% is due to special causes, such as operator error or machine malfunction.

---

v. The use of ranges to describe sample variance, rather than standard deviation, is a consequence of the computing power at the time the charts were developed. Today we could plot standard deviation with no difficulty.

Critical Input and Output Variables and System Design

The starting point for product stream design is to identify the output desired from the system, in the form of variables representing the characteristics of the required product (see Chapter 7). In addition, we have now explained that these variables are not fixed numbers, but statistical ranges of numbers, and we have defined a method for expressing the variation.

The characteristics of the required product are called critical output variables (COVs). COVs are a consequence of certain actions taken during manufacturing, beginning with the choice of materials used and their modification in the various process steps. The process actions, in turn, can be described as another set of variables, also in statistical distributions. These variables are called critical input variables (CIVs), and correlating them to the output is a complex intellectual task requiring detailed procedures in "cause and effect" logic. The best methodology developed so far for this task, which is also built into dynamic simulation models, is the fishbone diagram.[vi] This method works in steps, starting from the system output and examining immediate input variables. It then considers the input variables as an output of other process or management actions and continues to increasingly finer levels of detail until an adequate definition of all COVs and CIVs is achieved.

Figure 5.9 shows a fishbone diagram with most of the requirements needed to instruct a machine to transform a box of eggs into an omelet. We have chosen this example because it illustrates the difficulty of communicating concepts to a machine, even a particularly smart one.

vi. Also called Ishikawa diagrams, after their original founder.

Figure 5.9 - Making an Omelet with a Machine

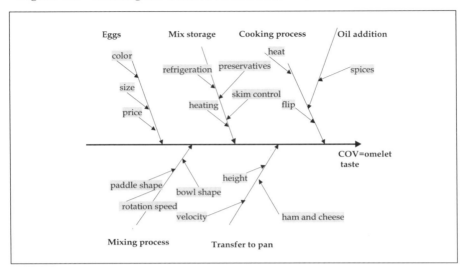

Once a person has been instructed how to cook an omelet, by a simple demonstration, he/she will most likely repeat the task in approximate manner and rapidly obtain acceptable results. A machine will need to digitalize the process in unequivocal instructions and this is why the diagram looks so complex; this may discourage us from building a robot for this particular application (a good decision!).

The Random Walk of Manufacturing Processes

Process control methodology is becoming the principal tool for the control of quality in manufacturing and its application. When COVs and CIVs have been properly identified, success is guaranteed.

Process control is also amenable to automation in the form of self-calibrating machines and instruments. The principle is that the machine continuously reads a prescribed set of signals, calculates sample positions on control charts, and notices any trends outside action of recalibration is carried out automatically, and if unsuccessful, a signal is sent to a human interface saying "Help!", "Stop!", or ringing loud bells. Error analyzing systems (EASs) are built into many control schemes and their purpose is to maintain variation within the control limits.

Thus, intelligent machines capable of adaptation can change the relative values of output and maintain calibration at the same time. This should not be confused with improvements in variation: the output of a manufacturing system may increase but variation will remain within the limits dictated by design.

Control charts are not incredibly helpful in making changes to a system's design because they average the variation in the system and hide the actual shape of statistical distribution. For this purpose, engineers and operating managers need to look at the shape of the statistical distribution and derive from this their design conclusions. The two curves in Figure 5.10 illustrate the point, using the cycle time of two stamping machines as an output variable.

Figure 5.10 - Two Processes in Control, One with Single Practice, the Other with Two Practices

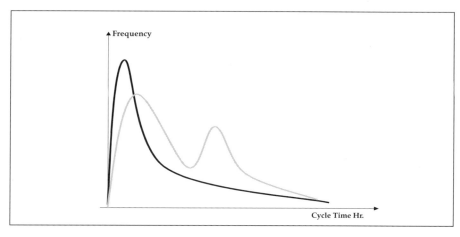

The two curves show the normal behavior of a constrained system; a machine with a certain operating practice cannot produce more

than its maximum design output, hence the skewed distribution. Plotted on control charts, both curves would be within their control limits and no special action would be suggested. However, looking at the shapes it is apparent that the black curve represents a single process, whereas the gray curve is the overlap of two different skewed curves belonging to different machine practices. This is a common example in manufacturing plants where different teams adopt variants of manufacturing methods on the same machine. Other common examples are the result of machine adaptation and require additional statistical analysis over and above control charts. With increasing automation, manufacturing systems are less transparent to the user and may require simulation in dynamic models to understand the underlying behavior.

> Good product stream management requires control in operation through the use of Shewhart methodology, and also requires an aggressive management of common variation causes. Common variation causes must be controlled by redesign over and above continuous improvement.

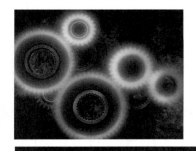

Chapter 6

The Statistical Meaning of
Profitable Manufacturing

Managers, Engineers, and Statistical Methods

The chapters on product streams and variation have shown that the performance of a manufacturing system can best be expressed as a set of statistical distributions, rather than a single set of numbers. For example, productivity measured as a number of parts per day will vary within a range depending on the complexity of the stream and the level of machine and process reliability. The higher the complexity and the lower the reliability, the broader the range of daily productivities will be. Similar considerations will be applicable to other business indicators such as cycle time, delivery performance, and cost per part.

This view contrasts the short-term management reports used at the plant or business division level: management reports are sets of numbers, not sets of statistical distributions.

There are some good reasons why variation is not portrayed in management reports. The selection of indicators is usually done to correspond with compliance reporting, not with real-time measures. Most compliance reporting variables are not directly connected with a process. A good example is the number of people employed each month, which is not an output variable of a product stream, but a management decision. Results are reported as averages (weekly or monthly reports), or at best, running averages (year-to-date reports). They are made easy to read even if the information is not complete. Business management is not accustomed to thinking in terms of input and output variables, and the two sets are often mixed together to provide a quick overview. In a typical income statement, for example, sales volume is a critical input variable (CIV), and cost of sales is a critical output variable (COV).

It is interesting that at the engineering level of machine performance, lean manufacturing methodology has introduced the discipline of control charts and statistical process control, but these methods have not been applied to the enterprise as a whole or even to the performance indicators of individual product streams. One of the reasons for not applying process control methodology is the difficulty of correlating engineering performance and economic indicators. Another, and possibly more important reason, is the difference in approach between engineers and managers; engineers are taught to deal with current events, whereas managers lean toward historic information and use intuition to project future trends. Both approaches have value and should ideally be combined to measure the potential performance gap of the enterprise against its ultimate capability and to observe the rate of improvement.

Some manufacturing experts advocate abandoning product stream economic measures and considering manufacturing plant cost as independent of mass flow because they are discouraged with the work done in activity-based accounting and product costing as measures of real-time performance. "Throughput" accounting has great value in continuous improvement because it allows the focus to be on overall plant cost and throughput, and on eliminating throughput constraints one after the other.[1]

A possible shortcoming of this approach is that continuous improvements are discovered and implemented by local teams and do not lead to the identification of the gap between the present state and the ultimate potential of the system.

Measuring Variation in Product Streams

We will now look at the statistical behavior of the two main drivers of manufacturing profitability: product price and product cost.

Statistical Distribution in Product Prices

Irwin Gross, professor emeritus at Penn State and founding director of the Institute for the Study of Business Markets, is one of the world authorities on value pricing and a passionate advocate of discovering the boundaries of customer value. According to Gross:

> Value Price is the hypothetical price for a supplier's offering at which a particular customer would be at overall economic break-even, relative to the best alternative available to the same customer, for performing a set of functions.[2]

Gross's definition suggests that the actual price a business can realize for a product will depend on customer perception and will vary within a certain range depending on circumstances. An example is the price for a dress during spring fashion sales in March or April compared with the price of the same item in the inventory sales at the end of the summer. It also suggests that a product capable of substitution provides an upper limit to achievable price.

A good example to illustrate Irwin Gross's point on value pricing is the U.S. Department of Transportation Federal Highways Administration's pilot project on highway tolls in San Diego, California. In order to control congestion in high-occupancy vehicle (HOV) priority lanes with minimal revenue loss, this project has established the prices that the customer is willing to pay at different times of the day before selecting alternatives (such as avoiding peak hours, using other routes or staying at home). The prices are set at a limit where some customers will choose to change commuting times, thus controlling congestion while at the same time ensuring full utilization of the lanes' peak capacity and revenue potential. The project provides for adaptive feedback during operation and prices will be adjusted to maintain correct traffic density, as shown in Figure 6.1.[3] The difference of $2 will induce a number of commuters to shift their travel time by one hour either way, but few will decide to go to work 2 hours earlier or come home 2 hours later to save another 50 cents.

Figure 6.1 - Value Pricing Pilot Project[4]

Maximum Toll Schedule (as of 3/99)
1-15 Value Pricing Pilot Project
San Diego, California

[PRIVATE] Maximum Toll	Morning Period (Southbound)							
	5:45 - 6:00	6:00 - 6:30	6:30 - 7:00	7:00 - 7:30	7:30 - 8:00	8:00 - 8:30	8:30 - 9:00	9:00 - 9:15
$4.00				X	X			
$3.00								
$2.50								
$2.00			X			X		
$1.50								
$1.00		X					X	
$0.75	X							X
$0.50								

[PRIVATE] Maximum Toll	Evening Period (Northbound)							
	3:00 - 3:30	3:30 - 4:00	4:00 - 4:30	4:30 - 5:00	5:00 - 5:30	5:30 - 6:00	6:00 0 6:30	6:30 - 7:00
$4.00				X	X			
$3.00								
$2.50								
$2.00		X	X			X		
$1.50								
$1.00							X	
$0.75	X							X
$0.50								

Looking at the Value Pricing Pilot Project as a business from the perspective of the Federal Highway Administration, this scheme optimizes the frequency of high revenues in the most desirable commuting times, and limits lower revenues to the off-peak hours, when the frequency of travel is reduced.

The price that the Federal Highway Administration business will be able to realize for its product (access to highway) can be represented as a skewed distribution where the peak represents the high-density users and the tail the very few users at 3:00 A.M. (Figure 6.2). Those who remember school math courses (or have read Chapter 5) will recognize this curve as an example of the "random walk paradigm," a skewed distribution common to any system that has an impenetrable boundary on one side.

Figure 6.2 - Statistical View of Pricing

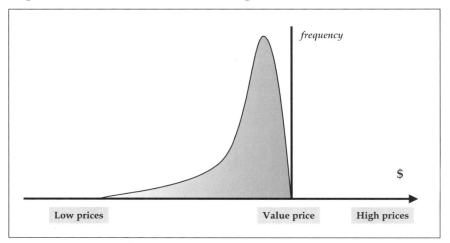

A business adopting value pricing will use a more cautious approach (since it is not a common objective to lose customers!) and will tend to price a safe distance from the boundary of substitution by a competitor offering.

Statistical Distribution in Product Costs

Product streams, as described in Chapter 3, are a representation of the flow of materials, cash, capital, and information in the manufacturing of a product. Through the use of product stream methodology it is possible to build static or dynamic models of any manufacturing operation and implement their findings in the establishment of a business. The same methods make it possible to look at the statistical behavior of flow in an ongoing operation.

An understanding of the basic relationships between product cost and the flow of products in a manufacturing system can be achieved without getting involved in the heavy math necessary for dynamic simulation. These relationships have to do with methods, not quantities, and are basic applications of industrial engineering concepts.[5]

Cost is an output variable; therefore it cannot be "managed" directly. The words "cost reduction" appear with unfailing regularity in manufacturing plan objectives and usually refer to using expedience to eliminate some cost, not redesigning the system to achieve lower cost. Addressing cost reduction directly is a North American attitude and is more a motivational approach than a recipe for successful implementation. European and Japanese firms believe in setting targets on the input variables that control cost. This is also the philosophy supported by Womack and Goldratt and built into the Six Sigma methodology used by GE, Motorola, and Allied Signal, among others.

To illustrate the relationship of cost to input variation we can look at the example of an operation that has already implemented sound manufacturing methods like the ones referred to above. This operation will have already eliminated most of the redundant cost factors and will be capable of measuring critical variables; hence, systemic variation will be more visible. If we were to select the case of a manufacturing operation lacking the discipline of manufacturing methodology, the frequent and random failures of the system would make analysis much more complex and less credible. How could you account, for example, for the cost of a team of specialists kept on staff to cope with the eventuality of computer network failure? The priority in this case is to achieve network reliability and eliminate the team, not to try to analyze variation.

In a product stream that is well designed and operated, cost is linked to input variable in two ways:

- Costs linked with mass (units of product flowing)
- Costs linked with time (manufacturing cycle)

As explained in Chapter 3, costs associated with units of mass are the costs of materials designed into the finished product and the cost of the cumulative yield of these materials. In mass flow the critical control variable for a manufacturing operation dealing with a prescribed product design is yield. Yield varies from product unit to product unit, within a range determined by process and machine capability.[6]

All other costs flowing through a stream are usually expressed as a function of time spent on each manufacturing step, time spent during machine setup, and time spent in intermediate storage.

This applies to variable costs, and allocation of fixed costs, and also to the utilization of the capital employed.

Even in a simple plant making one product, the cost of the product manufactured through the same product stream will still not be constant because cycle times on individual machines will vary with setup and machine reliability; internal logistics will have a range of performances; inventories will vary with customer delivery requirements and availability of transport; yield will vary from product unit to product unit.

Is it possible to identify the ultimate capability of a manufacturing system? The answer is yes, if we stay with the existing assets. Modeling and simulation combined with real-time measurements can provide an answer, showing that cost has a minimum boundary dictated by the design of the product and by the manufacturing system. The boundary is the extreme limit of the statistical distribution where all input variables are at their best. Although the boundary is intuitive, models provide a quantification of the minimum cost and also the probable statistical distribution, once the reliability of the system is known. The statistical distribution of costs (represented in Figure 6.3) is bound by a lower limit and is a skewed curve, mirror image in shape to that of price.

Figure 6.3 - Statistical Distribution of Cost

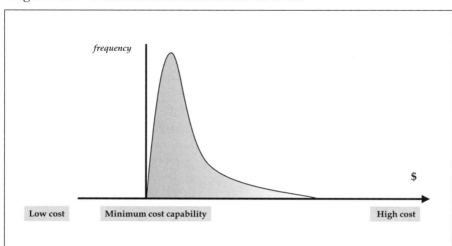

What Is Profitable Manufacturing?

We suggest the long-term viability of a business is dictated by the position of the two distributions of price and cost, which is why we think a statistical view of manufacturing business is necessary. We are referring to the overall business, not just the manufacturing processes in the plants; so far we have not found examples of manufacturing companies analyzing these measurements. Even the simple measure of overall cycle times and overall yields for a product stream rarely appear in management reports.

Figures 6.4 and 6.5 illustrate situations worth examining.

In Figure 6.4, we have a case where the distribution of costs is, most of the time, higher than the distribution of prices and the performance of the business will reflect this. The business may not be aware of the criticality of the two curves, but rather looks at the few positive events in the overlap of the two curves as a ray of hope for future success. These hopes would be ill founded since this is a systemically unprofitable business, which cannot improve without major change.

Figure 6.4 - Failing Business

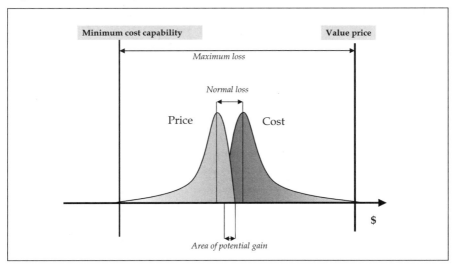

Figure 6.5 - Profitable Business

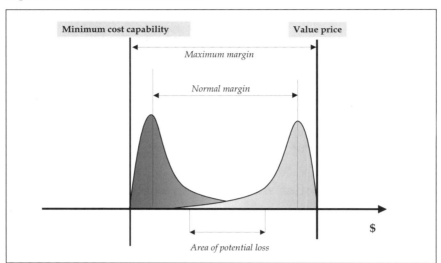

The business represented in Figure 6.5 is profitable most of the time since the distributions of cost and prices overlap only for a very small portion of the curve. A well-run manufacturing business using the right methodology will look similar. Understanding the two curves is still a very useful practice, because it will provide quantification to a number of strategic questions:

- Is it possible to improve the normal margin by either a different strategy of market segmentation or a different manufacturing configuration?

- If a new investment or an acquisition looks attractive, are these strategies going to improve the long-term position of the company?

> A change in the economic performance of a business results from changing the individual statistical distribution of cost and prices by narrowing the distributions or spreading the curves further apart.

We have talked about ideal cases where every operation is in control. Real-life curves in most operations are not as smooth as the one shown; they have lumps and discontinuities that are the result of either operations out of control or erratic pricing. The shape of the frequency curves is an indication of how controlled the business is. The closer to the random-walk distribution, the better the level of control. On the other hand, a normal or polynomial distribution suggests lack of success in controlling the business, because most occurrences of cost and price are far from the capability limits. Figure 6.6 shows this for manufacturing cost of a poorly operated plant.

Figure 6.6 - The Shape of Chaos

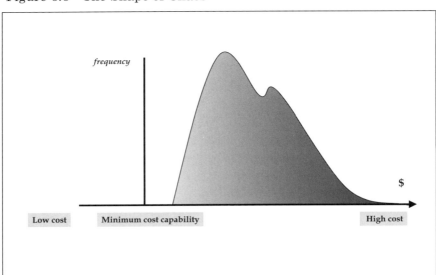

The shape suggests an overlap of different business practices in a chaotic fashion.

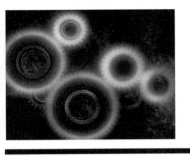

Chapter 7

In Praise of Simplicity

The Perplexing Green Lawn

> The basic principle of product stream design is to satisfy the customer requirements with the simplest solution possible.

Think of the apparently simple task of mowing the grass - an entire industry of equipment and service is built around it. Grass cutting is not a pleasant job; it is the source of untold misery on hot summer days as well as permanently impaired hearing and other occupational disabilities ranging from sprains to amputations. So why do we mow grass?

A neat lawn is esthetically pleasing and fits with the urban landscape. This idea comes from the image of verdant mountain pastures kept trimmed by grazing animals, but it is difficult to explain in an urban or suburban environment.

A neat lawn does not harbor untold pests and scratchy weeds, and a toddler can sit on the grass and not get those nasty red spots that result from sitting in wild grass. But why do we mow so many acres to make room for a small creature like a toddler?

We need to mow grass because it grows. The grass we are using is a traditional mixture of tall-growing species, which come from a time before lawn mowers, when lawns were occasionally trimmed with a hand tool but were otherwise left long and luxuriant.

So far, few solutions have come forward to solve the problem of obtaining a nice, neat lawn, free of maintenance. The most drastic solution is green polypropylene carpet, which is effective and long lasting. More organic answers may include low-growing grass species, genetically engineered crab grass available in a variety of tints, and other vegetable species with creeping habit.

Lawns, as we know them today, are a good example of a product that does not satisfy customer values, since nobody in his right mind wants to spend a good part of the weekend pushing a mower. So what does a lawn owner really value?

- The customer values a tidy yard with a soft flooring of a pleasing color, or a golf course with great fairways.

- A small segment of the market may be interested in lawn mowers, but this segment is rather specialized and we could cater to it separately with designer style machines capable of serious driving and possibly racing.

Lawns, and the industry surrounding the upkeep of lawns, are a classic case of market push, where value for the customer has not been built into the supply chain. An old, available solution has been promoted as viable, even though it does not conform to the customer's desires.

From the relatively simple issue of growing the correct plant, the industrialized world has built an entire lawn infrastructure with doubtful customer values as well as doubtful environmental credentials. To compensate for the incorrect solution, lawn maintenance is becoming more and more complex, with chemicals; signs warning of the chemicals; protective equipment; catalytic converters on two-stroke engines; automatic watering systems; and white hats to provide shelter from the ultraviolet rays of the sun while cutting lawns.

Cutting grass is a rich country paradigm - people in developing countries do not mow grass, they feed it to animals. One might speculate that as energy becomes less freely available in developed countries, lawnmowers will be high on the list of unnecessary gas gobblers and smog generators. Other solutions will have to be found for the upkeep of urban landscape.

Reflect on this example of lawnmowers. It is a common case of a product with limited survival potential, because:

- It does not provide customer value

- The supply chain is extremely complex and getting more complex to compensate for its inadequacies

More complexity does not necessarily make the business better. The industries behind lawn-mowing (seed growers, mower manufacturers, and garden services) struggle to make a living and will need to reinvent their mission. Some of the larger agricultural companies are working on new approaches: we hope they come soon!

An Approach to Simple Design

Product streams provide a tool to design the simplest supply chain that will satisfy customer requirements, without getting tangled in complexities (such as in the grass cutting industry). Designing for simplicity starts from these premises:

- Understand the customer requirements
- Decide the simplest supply chain that will satisfy the customer requirements with the minimum number of steps
- Use this chain and only this chain
- Control the quality of the product by controlling the variation of the process
- Ensure constant feed-forward and feedback of information through the supply chain

We will examine these items one by one in a quantitative way and discuss some of the techniques that have been developed to achieve results from the supply chain.

Understanding Customer Requirements

> The only way to fully understand the requirements of a customer is to talk to the customer.

This sounds trivial, but the reality is that in each step of the supply chain there are many assumptions of what is "good" for the customer; in other words, the supply chain tends to work in "push mode" rather than "pull mode."[1]

At each step of the supply chain the customer requirements have specific attributes such as:

- Product specifications
- Quality standards
- Order size

- Due date
- Acceptance standards
- Technical support
- Payment terms

... and a myriad of other details, all of which create the picture of what the customer really values.

A clear understanding of what a customer needs will result in correct inputs and correct design of a manufacturing product stream. Thus, a good design will not only satisfy the customer, but will also ensure that the order is processed correctly the first time.

In a supply chain, each process step is the "customer" of the previous process step: the requirements of the customer must be the driver of design for individual business processes.

When the "customer" is another part of the same organization, such as a department in the same company, it is more difficult to set supply chain priorities against departmental priorities. Here the leaders of the organization play a key role by creating reward systems that are tied to the success of the supply chain and not of the individual unit.

An interesting case in point is that of a metal supplier who has plants in several parts of Europe and supplies semifabricated materials from a central manufacturing plant both to its own national distribution centers and to third parties. This metal supplier found that customers of the national distribution centers received poor service; direct customers of the manufacturing plant always took priority. A new manager, well versed in supply chain management, solved the problem by inviting all customers to the same quarterly conference and making the supply chain information available to everybody. The local managers, aware that planning and scheduling of the manufacturing plant was no longer a plant issue and that manufacturing information was visible to customers, began to set priorities based on customer needs, and service improved rapidly.

> Inside a plant, product streams in discontinuous processes are a method of defining the manufacturing segment of supply chain management and adopting the same rigor in passing attributes and information from one machine to the next as a supply chain would do with different business organizations.

Attributes and information do not create a different organizational behavior: they are only a tool to communicate the details of customer requirements. The real challenges are in the organization structure and leadership style that are required to manage product streams.

Designing the Stream

The reason why product streams are not more broadly used in manufacturing when discontinuous processes are involved is the difficulty of designing them.

In the case of continuous processes, the stream is designed into the engineering of the plant, but what can one do in a discontinuous plant when a manufacturing step is followed by material movement to a warehouse, and three or four manufacturing alternatives are possible through similar machines?

Answers to this question have been developed in the Japanese and North American lean manufacturing plants. Figure 7.1 explains how this can be done, and can be summarized as:

> Decide the simplest route that will satisfy the customer requirements with the minimum number of steps.

Figure 7.1 - Design of a Product Stream

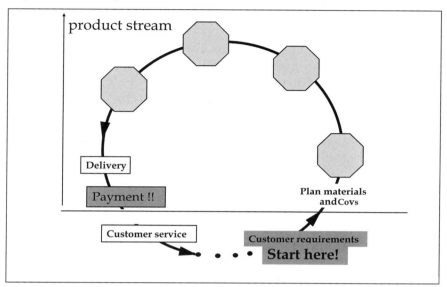

Beginning with customer requirements (step 1), it is possible to identify a preliminary material requirement from two separate sets of attributes:

- Qualitative attributes
- Quantitative attributes

Qualitative attributes determine which materials will ensure the correct end-product specification; they dictate what should flow through the product stream.

For an example, consider the product specifications for a blue dress. It is simpler to start from a blue cloth rather to start from a different color and introduce a dyeing step in the process. This is only true if the customer wants a commercially produced cloth. Should the customer ask for custom coloring such as tie-dyeing, the simplest product stream may start from another color since the dyeing process is unavoidable. This example illustrates how important the qualitative attributes of the product are. Pursuing this example further, other qualitative attributes will be the style of the dress, detailing of buttons, stitching, fit, and finish.

Failure to fully understand qualitative attributes is the reason for major rework and rejection of products in manufacturing industries.

Quantitative elements determine the flow through the product stream. This is not just mass flow but also the flow of other resources such as cost, people, and information. For example, if a customer has provided the qualitative attributes of the product, he/she will also advise you they need:

- Fifty blue dresses of similar construction but detailed in two different ways
- Delivery in 30 days
- Payment in 40 days after receipt

Qualitative and quantitative product attributes allow the design of a product stream. The challenge is to do this with the minimum number of steps.

To design systematically, the productive steps must first be identified and quantified. In this example, they are:

- Ordering of materials: 150 yards of blue silk, 50 yards of liner, 500 mother-of-pearl buttons, lace trimming for detailing one style, and sable collars for detailing the second style (3 days by UPS)
- Computer design of pattern (5 hours)
- Laser cutting of cloth to pattern (5 minutes per dress)
- Machine stitching of structural seams (10 minutes per dress)
- Hand stitching of visible seams and trimming (8 hours per dress)
- Packing and shipping (3 hours)
- Transportation (24 hours)

Between these productive steps are nonproductive steps when nothing happens, due to waiting for a machine, unplanned breakdowns, or a worker taking a coffee break. Nonproductive times are important since they can make up a large portion of the total cycle of making the 50 dresses. For example, a hand finisher will only produce approximately 80% of the time even if he/she works very hard. This had better be planned!

Should everything flow with no interruption, the numbers above will tell us that the first dress will be at the customer in approximately 5 days. To complete the last dress will take another 49 days if there is only one craftsman working 8 hours a day capable of finishing; this is past the delivery time required by the customer. This is a best-case scenario and in reality production time may double with nonproductive cycles.

The simplest product stream that would satisfy the customer quality requirements is one finishing craftsman. Hiring another one to reduce the finishing time is a serious decision. The consequences are:

- Introduction of a variation in the product stream in the form of a different style of stitching
- Complex contract negotiations (these types of people are considered precious and rare)
- Increase in cost

Should the shop renegotiate delivery with the customer? Should the shop try to finish some dresses by machine?

Use This Chain and Only This Chain

Renegotiating delivery may be viable, but the idea of finishing some dresses by machine is not. In doing so the shop is creating an alternative product stream, which has different quality characteristics, and will almost definitely result in an uncontrolled production.

Consider the case where a process stream has two alternatives at each step. For example, assume that the blue dress can be also cut by hand and finished by a machine (Figure 7.2).

Using two active processes at each process step gives us 2^n product streams, where n is the number of processes. It was difficult to understand all the nuances of the single stream, so controlling this level of complexity will be impossible!

Figure 7.2 - From One Simple Stream to Four

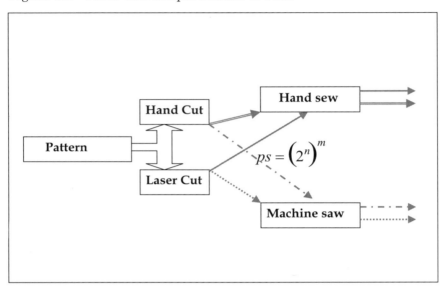

This is a simple case where the number of streams doubles at each step; in industrial processes not only are there more alternatives to each process step, but the sequence of these steps is also variable. If this were the case with the example above, we would have created $(2^n)^m$ product streams, where m is the number of alternative sequences (16 in this example). Most of the time the introduction of complexity in industrial processes is a shop floor decision, made in isolation: the consequences for the enterprise profitability are very large, as we will see.

Controlling the Quality of the Product by Controlling the Variation of the Process

The fundamental principle of lean manufacturing is to do things right the first time, and W. Edwards Deming has taught us that this can only be done by controlling the business process in all its aspects of information, people, and technology.

Blue silk dresses with fancy trimming are expensive. Even a volume copy of a well-known designer such as Armani or Anna Sui will cost over $1000. Defects are not acceptable! Neither is it acceptable to make 50 dresses and throw away 10 because they do not meet standards. The only acceptable method is to have reliable processes at each step of the stream.

Consider the stage where the blue silk dress is made and ready to ship: a well-designed product stream will immediately put the dress in soft paper, wrap it with breathable but waterproof material, and box it, before even considering storage. Minimum delay will ensure the absence of coffee stains or other accidental damage. A good process stream will also ship at regular intervals rather than waiting for all the dresses to be made, since this reduces the risk of accidental damage and improves cash flow.

It is through this kind of detail that control of the process is achieved. It is simply not enough to ensure that the cutting laser understands the tape from the computer; this is simple engineering control. Controlling a process means controlling every action that takes place in the manufacturing stream, whether it is a manufacturing step, or an inventory or accounting transaction.

Flow of Information through the Supply Chain

In the past 30 years control engineers have learned that the best way to run a technical process is to first predict by mathematical modeling and set the machine accordingly, then correct this continuously with information from the actual output of the machine (see Chapter 4).

In a supply chain, the flow of information in both directions is as critical as the control of machines. The design of a manufacturing product stream is only a preliminary pattern (feed-forward) and it cannot foresee the range of variability of the stream.

In the blue dress case, the dressmaker cannot meet one of the quantitative attributes: delivery. Feedback from the customer may tell him that the extended delivery is acceptable, provided dresses are shipped continuously as they are finished and the customer sees the manufacturing progress "on-line" through a network. This is a different product stream, which is now rooted in information technology.

Information technology, networks, and e-commerce will ensure that information continues to flow through the stream in both directions during manufacturing: there is no reason that the same approach cannot be taken, without computers, by simply talking to people. It is only a question of the right business practice.

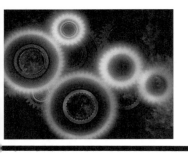

Chapter 8

Organizations for Real-Time Decision Making

The Importance of Decision Making

In describing the principles of product streams, we have stressed the need to manage all aspects of product flow, starting from two basic principles:

- Always flow the right product: that is, the product that a customer wants at the end of the supply chain.

- Once the product is correctly defined in all its attributes and process settings,[i] flow must proceed rapidly and without interruption.

In real life continuous flow does not exist. The closest approximation to continuous flow is found in processing plants where all the manufacturing steps are physically linked together (for example, a pipeline or a water purification plant). In all other cases, which include batch processing operations and discontinuous manufacturing plants, the active process steps may represent a small portion of the total cycle. Nonproductive steps make up the rest of the cycle and, in most manufacturing businesses, are controlled by people, not machines.

As an example, think of a polymer intake manifold for a car, which is fabricated in separate steps as shown in Figure 8.1.

Figure 8.1 - Active Cycle Time in a Polymer Manifold Manufacture

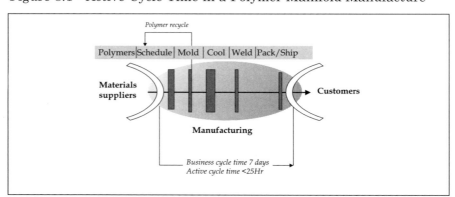

i. For a more specific explanation refer to the blue dress example in Chapter 7.

The active part of the manufacturing that is machine controlled consists of ordering the materials and scheduling the plant (12 hours of e-business), injection molding the components (10 minutes), inspecting and cooling (10 hours), vibration-welding (3 minutes), and packing and invoicing (2 hours). The remaining 6 days of the cycle are spent in nonactive steps that are not machine controlled and are dependent on human decisions. It is not suggested that this time is unnecessary, as it may be required by machine setup, economic batch size, or planned inventories in the supply chain. The critical point is that people control this time and its management requires immediate action at the point where events take place: the customer's plant, the factory floor, the design office, the maintenance shop.

In discussing the role of computers, we have given examples where automating predictable events can reduce the amount of decision making in manufacturing, but many events in the real world are not in the predictable category. Nonpredictable events require skilled decisions, either of a strategic nature in the general direction of the business or of a tactical nature at the manufacturing operation level.

In manufacturing, the majority of decisions that affect the performance of a business are at the execution level, not at the management level. As computer-based decision making becomes more sophisticated we suggest that the first application will be the replacement of middle-management staff. It is possible, for example, to conceive of an intelligent machine replacing a planning or operations manager with positive results. It is hard to conceive of a machine replacing the technician who keeps this "mechanical manager" operating smoothly: self-tending machines are a long way away and probably uneconomical.

We have represented this dependence on human decisions in Figure 8.2, using the analogy of a quarry, where the real action happens with dynamite sticks at "the rock face." Applied to manufacturing, this picture is meant to illustrate that the majority of effort is in the execution, a task in which manufacturing organizations must excel. What distinguishes an excellent business from a mediocre one is not the ability to make decisions, which is common to all businesses, but the quality and the timing of the decisions.

Figure 8.2 - The Rock Face

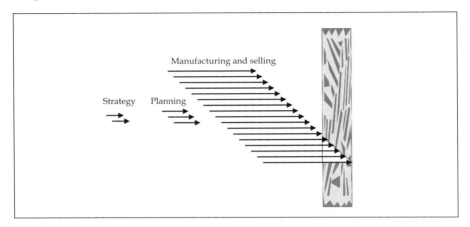

Organizational Structures

The decision-making process in a business is strongly dependent on organizational structure and the style of management. Decisions need to be made rapidly as well as correctly, which implies the necessity of an appropriate level of knowledge and skill.

To summarize in a simple statement the requirements of an effective manufacturing business, using a product stream approach,

> all decisions should be made at the point closest to the execution where the necessary skill exists.

We can now examine the ability of different structural organizations to fulfill this requirement. Traditionally, organizations have been structured in two ways: functional organizations and project- or product-oriented organizations. With respect to decision making, these two structures allow for either quicker or more informed decisions, but not both because of the inherent limitations on the type of interactions the employees experience.

Functional Organizations

In the functional organization, employees are divided into departments defined by specific skills: accountants with accountants, operators with operators, and engineers with engineers.[1]

Functional organizations have evolved from a traditional hierarchical style of management, which has ancient roots in armies, states, and religions and is natural to all highly evolved species.[2] This could be the reason that a functional structure "just feels right" to most people. It offers a clear sense of belonging and allows for structured competition, including the challenge for leadership directed at the "old bulls of the herd."[ii]

Hierarchical management styles work well in small businesses where an organizational structure is not necessary: the boss makes all the decisions and implements them with some help from employees. The result is informed, fast decisions and quick execution.

For example, Henry Ford made all the decisions in the Ford manufacturing organization, right down to the smallest detail (there was not one bolt in the Model T that was not approved by Henry). Yet his style was not one of a dictator; on the contrary, he had a strong paternalistic attitude toward his business and introduced many social innovations that were unique at the time. One limitation was that this single-man decision-making process ensured that change could only happen at the pace that Henry Ford could muster. The death of the Model T came rapidly and painfully when competition designed and developed improved systems for transmission and suspension. Ford could not react fast enough, and in 1927 the last Model T rolled off the Rouge Plant assembly line.

---

ii. Watch carefully the behavior of cattle in a pasture and you will see the same patterns that characterize civilized functional societies, without the niceties!

Facing the same issues as Henry Ford, many mass production manufacturing businesses have sought to rectify the situation by adding strong staff support for the hierarchy, and have thus developed functional organizations as described in Figure 8.3.

Figure 8.3 – Functional Organizational Structure

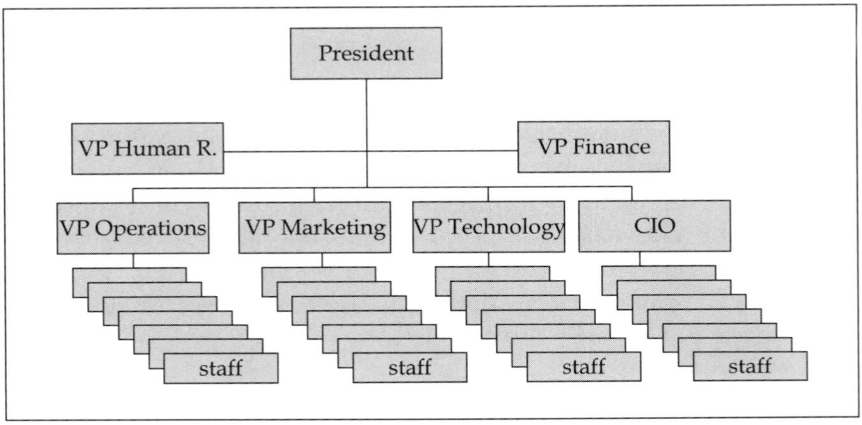

Whereas a purely hierarchical organization has great value in quick and informed decision making within the span of control of one person, the same cannot be said for functional structures. This is particularly true when some of the skills are acquired externally, as is the case with many consulting services. The quality of the information is often excellent, but the time required for decision making can prove fatal in a quick-moving market.

In the 1970s and 1980s, functional organizations promoted the growth of a new maverick, the "knowledge worker," which played a dominant role in the Western manufacturing industry.[3] Well-remunerated, intellectually bright, and often highly opinionated, knowledge workers not only reshaped the thinking of many companies, but also developed a blossoming consulting industry. In selling their services, inside or outside the organization, knowledge workers inadvertently promoted a climate where no decisions could be made without a "study" and reinforced the major disadvantage of functional organizations: speed in decision making.

This points to two serious pitfalls of functional structures, even with the best of specialized support:

- Leadership can become diffused and accountability lost.

- Machines only understand simple direct commands, which only technicians can construct. These messages need to be technically correct and consistent with business strategy at the same time.

Regarding the first problem, Susan Rivard has observed many organizations where the division of tasks between the planners and the doers becomes institutionalized as an accepted method of operation.[4] The leaders of the organization become so dependent on the implementers that they abdicate leadership, often without realizing it. How many times have you heard a senior manger say, "I don't understand what this process does, but I have a wizard of an engineer who knows all about it!"

This senior manager is perfectly correct. A single person may not be able to span all the details of the business process, but the organization is also aware of the shortcomings of the leader and starts to appoint unofficial "gurus," who provide the missing knowledge without a shared vision of the company goals. The functional organization may then be better represented as in Figure 8.4.

Figure 8.4 - A Different Representation of a Functional Organization

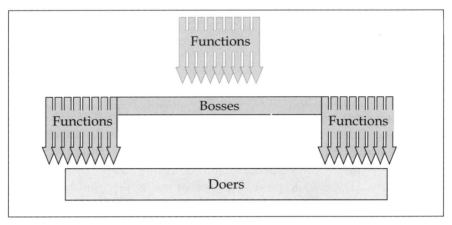

Concerning the question of communicating with machines, they may be becoming smarter, but they only understand single direct

commands. Strategic messages are too complex and must be translated into a corresponding technical language.

At a U.S. metal manufacturing plant, a large and expensive rolling mill has adaptive control systems that allow the machine to optimize gauge control and mass flow. The optimization has specific objectives, which are part of the company goals and relate to the end product, and not to the performance of a single machine in the supply chain. Inserting a program in the supervisory computer does the "talking" to the mill, and only a few engineers in the plant know the specialized code. One enterprising engineer had a better idea for optimization, one far superior to the simple rules agreed upon with the plant manager. During the night shift, he inserted different command software, which he had personally developed to increase machine output. Unaware of any priorities, the machine followed the control engineer's instructions and strange product variances showed up in the customer plants. Nobody ever traced the well-intentioned yet misguided functional intervention until years later when the mill control module was updated and the program deleted.

The moral is simple:

> In a manufacturing business the leader must know and control the business process and not abdicate to functional intervention.

The alternative is analogous to driving a car, a task that requires single-minded management, with constant running instructions coming from an engineer, a lawyer, and an accountant. A functional organization controlled by a single leader works well for businesses that can be spanned by a few individuals. Very few businesses today will admit to having a functional organization with a hierarchical style. This is a pity. For some businesses, it is the correct form of structure, particularly at the top of the corporation. If we look past the examples where leadership has been abandoned in favor of advice, the value of informed decisions is very real, and enlightened strategists have created great wealth in the service sector. The revolution of banking systems, new investment services, global communications, and mass tourism all have great examples of organizations driven by knowledgeable hierarchical style leaders.

The larger question in manufacturing is whether functional organizations are compatible with a "lean" style of management. Functional organizations do not have to be inefficient and bloated with redundant jobs. There are many examples of lean head offices that perform well the few tasks required for shaping the direction of the company.

It is somewhat amusing that the "we are a team" slogans may lead businesses to camouflage their real structures through informal dress codes and friendly environments. Transparency is much more reassuring to employees than friendly atmosphere, as some businesses are discovering with "open book management." This practice is being successfully implemented in formal organizations such as professional partnerships, as well as in informal new-tech businesses.

A lean hierarchical leadership at the top of the company, leading strategy and company direction in execution, not only is desirable but is a necessity in most traditional manufacturing plants. This top structure includes few people and does not preclude a completely different type of organization at the "rock face." We agree that functional organizations are not compatible with supply chain management and product streams, but these are concepts that apply at the execution level and require a different structure.

Product-Oriented and Project-Oriented Organizations

In a product-oriented or project-oriented organization, employees are organized in accordance to the content of their work, not the function they perform. Examples of project-oriented organizations are seen in large engineering firms contracting projects such as the construction of manufacturing plants and power utilities. Product-oriented organizations are common in the manufacture of consumer products.[iii]

---

iii. The classic case, studied in all business schools, is the Minnesota Mining and Machines Corporation (3M).

Since the two concepts are similar, we will explore the product-oriented organization and its values. In a product-oriented organization each unit consists of people with different backgrounds and functions (as shown in Figure 8.5) and the size of the unit may vary from a small team to a large organization. Examples of small teams are seen in "high-tech" manufacturing when launching a new product.

These are exciting environments in which to work and the management style can be informal and personally oriented. Large units are seen in established product divisions of automotive companies, machinery manufacturers, and consumer durables and are usually associated with a brand name.[iv]

Product-oriented organizations are highly compatible with supply chain management and product streams. They also lead themselves very well to open management styles and to easy communication with customers.

The most important feature is their ability to make decisions quickly and maintain their focus on customers. As we have seen, this is an essential feature of flowing products through product streams.

Figure 8.5 - Product-Oriented Organization

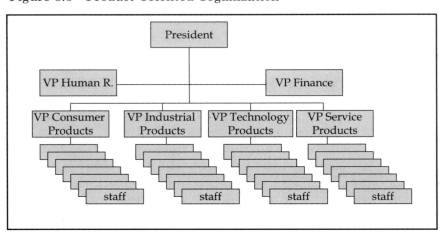

---

iv. When the units become very large, as in the product divisions of large auto manufactures, they become closer to a functional organization.

Do these organizations still have all the information they need to make high-quality decisions quickly? Womack and Jones make an interesting point in their paper "From Lean Production to Lean Enterprise":

> Functions do much more than accumulate knowledge; they teach that knowledge to those who identify their careers with the function, and they search continually for new knowledge. In the so-called learning organization, functions are where learning is collected, systematized, and deployed. Functions, therefore, need a secure place in any organization.[5]

Product-oriented organizations have more difficulty in fostering learning, particularly the practical learning that happens between those of similar functions. In other words, they offer a poor environment in which to create specialized excellence. This may be essential to the competitiveness of the business.

Matrix Organizations

The answer to the problem of the functional and product-oriented organizations may seem obvious: blend the two and exploit the good points of each one. This "matrix organization" has been developed and implemented in some companies. One of the most sophisticated global implementations of a matrix is found in the automotive sector and is literally a cross between the functional organization and the product-oriented organization (Figure 8.6).

Figure 8.6 – Matrix Organization

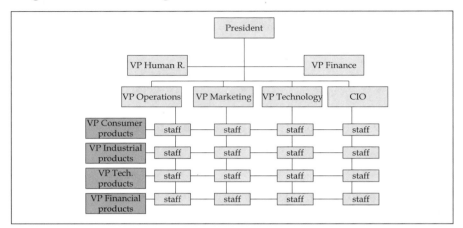

The matrix organization allows for the quick transfer of information by allowing employees to be part of two groups. They are part of a functional group so they can have contact with those who have similar professional skills, and they are also part of a product group, where they have access to employees in other functions working on the same product stream. This should allow for both quick and informed decisions at the same time.

Matrix organizations have been around since the 1960s, with little or no impact on the performance of the manufacturing industry. The lack of success is not inherent in the structure but in the style of management prevalent in Western manufacturing. Matrix organizations simply cannot work with a hierarchical management style or overstaffed functional groups. The reasons are again connected to basic human instincts of competitiveness and leadership challenge, which are promoted in the matrix by the assignment of double accountabilities to functional managers and product line managers. The effort expended in organizational maintenance (including individualized leadership training and team performance training) far outweighs the advantage of the matrix structure and consumes excessive energy in inward-facing activities.

Progress has been made recently in making matrices work with a different management style. This "self-directed team" is another of Toyota's ideas. The concept is simple: "Take your best resources and organize them in teams with all the necessary skills, then give them full accountability for decisions in their unit."[6]

The reason why the self-directed team allows the matrix structure to work is the removal of control in both functional and product-oriented organizations and the assignment of accountabilities "at the rock face." This process eliminates a large portion of the middle management and focuses the senior staff on strategic needs: the product manager on the market, the engineering specialist on creating engineering skills, and the production manager on running operations smoothly.

Whenever the team management style has been implemented successfully, the results in productivity and customer satisfaction have been dramatic. The main ingredients of success are:[7]

- Clarity of common objectives and goals
- Ability to communicate directly and honestly
- Mutual respect of skills
- Ability to release control

According to Arlette Bouzon, adopting the format of team approach, but not the intent, is a commonly observed practice and renders the matrix inferior to hierarchical organizations.[8] Bouzon has observed partial European solutions where accountability is not released and demonstrates the incompatibility of matrices with hierarchical styles of management. The result is conflicting instructions, indecision, and demotivation.

Understanding how to work in a team takes training and practice. It is not an instinctive trait; it is an acquired skill. To agree that teamwork management styles are desirable is easy for any rational person. To apply this style day after day in situations that often call for difficult decisions requires a level of discipline and training that very few organizations manage to develop.

A particular set of successful organizations employ "open-book management" to reinforce team behavior. This is a management style that allows all employees in a team access to goals and performance indicators of the business that are usually the prerogative of upper management, even in the best team-oriented matrices. By keeping employees informed and educated on broader business aspects, these companies are able to provide more scope to their work and thus increase the motivation and awareness required for sustaining the discipline of daily problem solving.

Rewards and Challenges

A matrix organization structure with a team management style appears to solve all manufacturing industry problems forever. Not quite!

To present an objective picture of these organizations, we also need to mention the opponents of the movement. Schenk and Anderson argue:

> While the social and health problems associated with Fordism (mass-production) are indisputable, the ability of lean-production to correct these ills has not been shown. In fact, there is a growing body of literature that raises serious doubts about the alleged positive impact of lean production on working life.[9]

These authors are questioning the ability of humans to sustain the emotional and physical demands of team-based matrices for prolonged periods of time, in spite of the team ability to make timely and knowledgeable decisions.

Is it realistic to expect employees to complete their daily tasks in addition to keeping in touch with members of adjacent teams in the product stream and be aware of what is going on in other aspects of the plant? Even more important, is it possible to invest the energy in the continuous training required to learn new techniques?

A survey of Canadian auto-manufacturing employees has produced results like the ones shown in Figure 8.7. These statistics show that the majority of workers in Canadian automotive manufacturing plants feel that they do not have time to learn other jobs, and that their traditionally assigned tasks keep them busy enough during their scheduled work hours.

Figure 8.7- Workers Responding That They Have Little or No Time during a Job Cycle to Do Things Other Than the Assigned Tasks (%)

	All direct production	Male direct production	Female direct production
Chrysler	68.51	66.01	79.63
GM	84.47	84.42	85.54
Ford	67.23	66.37	85.71
CAMI	64.71	68.18	55.56
Total	75.45	74.82	80.27

This could be a transition stage, but significantly more time has to be invested in the design of teamwork if we wish it to work consistently in a Western society. The expectations of a matrix organization go far beyond the allotted work week. They assume a level of dedication from the employees that may or may not be practical in the long run, no matter what rewards are offered.

We believe that what the auto workers are experiencing is one facet of a whole society in transition: the real drive behind the stressful environment is the enormous capability of the new machines and of the global networks, which we are struggling to harness.

Go to an airport on a Sunday night or at five in the morning, any day of the week, and you will see the "laptop commuters" trying to cope with global business. Until machines become capable of taking on more tasks in automatic communication in supply chain management and in manufacturing operations, these commuters will find the management of their personal time and energy an increasingly demanding task.

On Human Frailty

To explore a little further the human implications of the structures we have described, we will look at the case of Mary and Bill (real people with altered identities, in our circle of acquaintances).

Mary is a highly qualified investment fund manager working for a large international corporation. She is a well-rewarded senior executive in her company, with a strong financial incentive tied to the performance of her fund. At the age of 34 she already has a well-rounded portfolio of personal investments as well as a significant contribution in two pension plans.

Mary's partner, Bill, is a professional engineer consulting in telecommunications, through a Web site. He has a 50% share in his business; the rest is with two other investors who are members of the company board of directors. Bill has no direct employees, but his contracts involve assembling problem-solving teams from all over the world. Bill's income is highly variable, depending on major contracts, but has a large upside potential.

Figure 8.8 - Explicit Business Expectations for Bill and Mary

	MARY	**BILL**
Working hours	*8 hours/ 5 days* *1 hour commuting / 5 days*	*Variable / 7 days*
Training	*Company team work* *Professional upgrading*	*Professional upgrading* *education and research*
Personal image	*Business formal dress code*	*None*
Supervision	*Managed by objectives*	*Board of directors*

These two individuals work in what we described as matrix organizational systems with a team management style: Mary works in a real one, Bill in a virtual one. Their acceptance in their workplace is subject to certain expectations, some explicitly spelled out, some implicit if they desire to be successful. We have summarized the principal explicit expectations in Figure 8.8.

Mary and Bill are both very successful and their progress is highly dependent on understanding the implicit expectations in their respective environment.

Mary is normally in the office at 7:00 A.M. and rarely leaves before 5:30 P.M., thus doing most of her routine work outside normal hours. She must remain free to attend to customers and consult with colleagues during the regular workday and manages to exercise at lunchtime three times a week. Lunches, dinners, and golf games with customers are expected during the week and on weekends, on a regular basis. Her manner with customers must be pleasant and relaxed, but inspire professional confidence and demonstrate a willingness to solve problems. Mary knows her supervisor well and the two communicate easily, seeking mutual advice and discussing problem areas, which avoids any surprises at review time. There are challenges in Mary's life: the management of her personal time, which is rarely successful, and the constant discipline of maintaining the correct image in front of customers and colleagues.

Bill has a different challenge: he is free to work 24 hours a day or not at all. The feedback of the business he has is not immediate. Professionally, he is on the leading edge of technology, but he must exercise discipline in selecting customer accounts and evaluate quickly where the best rewards may be. At the same time, he must maintain open relations with all potential accounts, participate in trade shows, and promote his products. Relationships with the board of directors require social time and close communications. Bill has little control of his personal time and the 7 days per week accessibility to customers is a strain. Figure 8.9 summarizes some implicit expectations of the jobs of these two individuals.

Figure 8.9 - Implicit Business Expectations for Bill and Mary

	Mary	Bill
Working hours	Accessible to customers and colleagues. Heavy social schedule	Accessible to customers, 24 hours a day for seven days a week
Training	Competitive knowledge base	Leading edge technology
Supervision	Broad base of inputs from colleagues. Expected to lead in her field at all levels of the organization	Strategic leadership of business
Personal image	Confidence inspiring, professional approach. Healthy and balanced	Confidence-inspiring. High tech image

These two young people are a current example, at the upper end of demands and rewards, of what a global business organized in a matrix form is likely to expect. Similar demands will be made on other employees to a greater or lesser extent, the most important of which are self-management and initiative. This is much more stressful than following orders and makes heavy physical and emotional demands on the individual.

Are the demands sustainable? Does a level of reward exist that will make them acceptable? The answer to these questions is not apparent.

The Industrial Revolution replaced hard manual labor with mechanical devices; the information revolution must replace travel, meetings, and complex paper communications with intelligent machines and better working methods.

> We may be in a learning period during which we are duplicating a lot of functions that machines can perform, and it may take some time before we learn the real roles of humans and machines.

Glimpses of the Future

Kevin Warwick is a cyborg.[v] This intense and driven professor of cybernetics at Reading University has had a chip implanted in his arm which connects to his nervous system and send signals to a computer, performing magical acts such as welcoming him to the lab and opening a door and offering him a cup of coffee. He now plans to test progressive feedbacks to his brain, and his wife has agreed to get an implant as well: the metaphor of sharing feelings may be closer to reality than we think.

This may not be directly relevant to manufacturing, but an important point has been made - man and machine are becoming closer together, creating a mutualistic relationship through easier interfaces. We do not need to go as far as Kevin Warwick to think of easy access by voice command to machines and networks, questioning information systems, problem solving accomplished without the ponderous work of examining control charts, or eventually even becoming trained through multisensory immersion in a virtual environment.[vi]

Even today there are many real examples of businesses that cannot exist without this mutualistic relationship with machines. These include all e-business, electronic news media, and most financial transactions. (Bill, our high-tech consultant, is not implanted yet, but his business exists because of the machine.)

Cyborgs, if they existed, would communicate easily and alleviate many of the demands of the present business organizations (Figure 8.10).

---

v. A cyborg is a mutualistic relation of man and machine.
vi. For example a virtual trip to a foreign country with interactive programs will teach a language in record time, if you are willing to play the game.

Figure 8.10 - Cyborgs

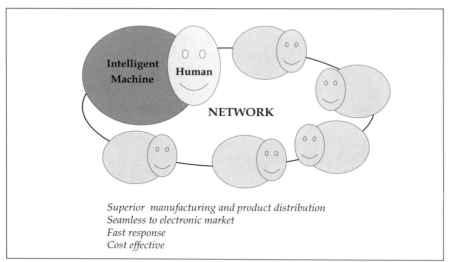

Superior manufacturing and product distribution
Seamless to electronic market
Fast response
Cost effective

At present there is no expectation that many consumer goods will ever be manufactured exclusively by cyborgs. Nonetheless, we can expect that the level of automation will increase to the stage where all decisions will be computer-assisted, both in manufacturing and in customer service.

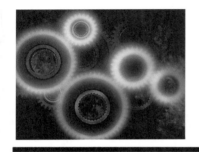

Chapter 9

Next-Generation
Manufacturing and the
Virtual Reality Plant

The Next-Generation Manufacturing Project

The buoyant economic environment of the 1990s shifted the attention of investors and the public alike from traditional manufacturing industries into the world of new technology. There is reason to believe that in the future, new fields of technology such as bioengineering, global networks communications, nanomechanics, and virtual reality will create businesses with superior growth potential.

Perhaps more importantly, these new businesses will change the way most of us live, by selling a service rather than a physical good. As a consequence, the monthly budget of a family 50 years from now may look substantially different from today, with increased allocations to communications services and remote learning, increased use of smart machines at home, and a reduction in transportation needs.

The road leading to these changes is likely to be punctuated with failures and disillusionment. However, we are on the road to a major shift toward less intrusive and increasingly user-friendly machines. A person living in 2050 may regard the most efficient car designed in the past decade in much the same way we smile at the cumbersome and wasteful mechanical monsters of the Victorian era.

These projections may be positive in many ways, but our physical needs must still be met with durable goods such as bathtubs, water systems, power distribution, food, and drink. How is this going to happen if manufacturing does not become a more attractive business? Having goods manufactured in China is only a temporary solution. Soon manufacturing will no longer be attractive in developing economies. The only answer is a radical change in productivity and resource efficiency.

In 1995, a group of industrial leaders and academics set out to answer the questions on the future of manufacturing. In 1997 they published a report entitled "Next Generation Manufacturing: A Framework for Action" (NGM).[1] The Next-Generation

Manufacturing Project set out to identify the competitive drivers in the future business environment and define the attributes required to compete successfully in the global market of the twenty-first century. The conclusion of the study was a plan of action for industry, government, and academia. Figure 9.1 summarizes the strategic direction and alignment of what the report calls a Next Generation Company.[2]

In Figure 9.1, the NGM Company is described as an integrated entity of people, business processes, and technology, with excellent response capability. This responsiveness applies to:

- Customers
- Plant and equipment
- Human resources
- Global market
- Practices and cultures

Figure 9.1 - The Next-Generation Manufacturing Company

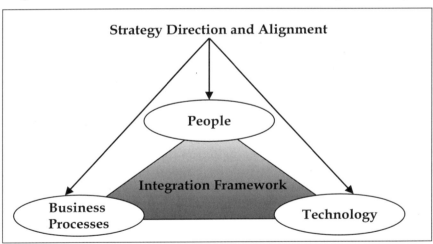

Since the NGM report was produced, many of its ideas relating to market responsiveness and global business have become reality in industrialized nations. Much less has happened in technology and manufacturing plant integration.

The recommendation of NGM is that:

> manufacturing must be addressed as a total, dynamic system that tightly integrates people, processes, and technology.

This is much more complex than continuous improvement on a local scale and presupposes a vision of product streams and their dynamic response.

Outside manufacturing, the past 5 years have seen excellent examples of integrated systems and supply chains. One example is in the field of distribution and service. Many distribution businesses are new and have been created as integrated systems. Manufacturing plants, in contrast, are typically designed as stand-alone entities that may later grow to be part of a system. This could explain the barriers encountered in manufacturing when attempting to change the culture toward system responsiveness, as described in NGM.

Redesigning the Manufacturing Enterprise

Tire manufacturing is a classic mass-production system complete with mile-long assembly lines, dedicated workstations performing specialized tasks, and off-line quality control systems. Lean manufacturing can be discussed with any of the large tire manufacturers and they will say that they are practicing it; witness the quick deliveries they can offer out of finished goods inventories. Large tire plants were never designed for lean manufacturing or teamwork. Attempting to adapt them to the demands of lean supply chains may respond to a need for the lean image, but does not create a real competitive advantage.

Tire making is a secretive business tied to expensive assets. The risks involved in any technological changes are enormous because of potential liabilities. This discourages any step changes that could revolutionize the industry.

In 2000 Bridgestone/Firestone Inc. was forced into a massive tire recall. The origin of this recall may or may not be traceable to technical defects generated in a conventional tire-manufacturing line. This incident has now cast doubt on the reliability of existing operations and their heavy dependence on human decisions. Tire makers now want to reinvent their industry with a recipe similar to the steel mini-mills of the early 1990s.[3]

The Italian company Pirelli SPA (manufacturers and distributors of special cables and wires, and photonic products) opened a mini-plant in 2000 built around a compact and totally robotic system where humans have no contact with the product. France's Group Michelin tire company has a similar version installed at eight plants around the world in conjunction with conventional lines. US-owned Goodyear Tire and Rubber Company is working on a mini-plant a fraction of the size of a conventional tire line and capable of being totally automated. These are all signs of the tire industry moving toward dedicated mini-plants (possibly located in proximity to auto plants or tire distribution centers) to significantly improve their economics.[4]

The expectation of change in a mass production system such as tire manufacturing has to be tempered by the existence of a massive infrastructure as well as the risks involved in the change. Also, improved economics will not necessarily result in improved margins since some of the gains will have to be passed along the supply chain. All this points to a gradual conversion in a longer time frame, rather than a revolution. Change is inevitable since staying with the existing system has proved to be risky in addition to being uncompetitive over the long term.

The tire industry is not the only example where the most likely future is the fragmentation of large manufacturing facilities into small agile units capable of the responsiveness advocated by NGM. The whole supply chain for the automotive industry is moving toward highly automated small factories located close to the supply point.

Outside the automotive field many packaged plants are fulfilling the role once performed by massive industrial complexes. Consider the example of Air Liquide, a $5 billion supplier of industrial and health-care gases. One of their products is nitrogen, once generated by distillation in large plants and then distributed in liquid form by truck. In 1999, 3000 micro-plants based on membrane technology supplied industrial-grade nitrogen to 35% of Air Liquide customers on their site. These micro-plants are tailored to the volume required by each customer and are totally automatic, requiring only periodic maintenance by visiting Air Liquide technologists.[5]

In the food industry, packagers of soda pop also operate totally automated filling plants close to supply locations. In building products, plastic and concrete components are being created every day in small, specialized plants.

The implications for a manufacturing industry stepping away from central manufacturing facilities are complex. While theoretically it may seem like the correct solution, it is not apparent that the level of reliability offered by smart machines allows operation with no human intervention. On the contrary, when intervention is needed it usually requires high technical knowledge from the operating technologist and a micro-plant may not be able to afford such personnel. The same can be said for product changes, and even minor retooling.

The decision to fragment manufacturing is dependent on the technological capability of the system. In the case of the Air Liquide plants, the capability was proven over many years of operation. Liquid nitrogen is produced with the kind of reliability expected out of home refrigeration. This may or may not be the case with tire mini-plants since the physics of rubber polymerization are much less understood than the physics of compressing gases. Fully automated tire plants embodying the best learning machines may still fall short of humans' intuitive sense and produce unreliable results for different reasons.

For all these reasons we may expect large-scale manufacturing with some of the elements of lean production to exist in parallel with small, agile, fully automated plants for a long time to come.

Understanding the Gap

The ability to make the right decisions in restructuring a manufacturing business is critical not only in content, but also in time. This makes an analytical approach in support of the decision process more complex, but not unapproachable with present computing capability.

To illustrate this we can use an engineering technique called technical gap measurement and apply it to the cases of the tire mini-plant and the liquid nitrogen plant.

Measuring Technical Gaps

> Technical gap is the difference between the design specification of a product or a machine system and its actual performance described statistically.

The process of making a tire involves the preparation of several rubber blends, to be applied to different layers of the tire.[i] Then the tire carcass is constructed over a mold. This latter process is carried out for the most part by skilled operators (tire makers) who are assisted by machines and responsible for performing the correct sequence of operations and for applying the correct ingredients. The completed carcass is then cured in a mold and later trimmed. The uncured tire is very sensitive to damage and dirt contamination if carelessly handled.

The technical field of predicting the performance of a finished tire under different driving conditions is dominated by experimental evidence rather than scientific prediction, and the know-how is closely guarded. However, progress is being made. Special rheometers can assist in predicting the properties of a particular batch of ingredients after curing. As well, the stress behavior of a finished tire (a composite product made out of many layers) can be modeled by finite element analysis and that model can be adapted to the results of physical tests.

---

i. In the average passenger tire, which weighs about 11 kg, carbon black accounts for 28% of the weight, synthetic rubber 27%, fabric 16%, steel 15%, and natural rubber 14%.

Comparing the complex process of making a tire (which involves a high degree of uncertainty in the process and the properties of the finished product) with the highly predictable process of making liquid nitrogen, we find two different manufacturing operations. The ideal process of making liquid nitrogen is easily modeled from physical laws. Selecting critical output variables (COVs) and comparing them with the statistical performance of the plant allows for easy measurement of the technical gap. In Figure 9.2 we have chosen energy requirement as the measure, and the solid line represents the theoretical energy required. The frequency distribution represents actual measurements in daily operations, which are always greater than the theoretical limit.[ii] The technical gap can be measured by describing the distance between the theoretical value and the parameters of the distribution curve (maximum, minimum, and mode), as shown in Figure 9.2.

Figure 9.2 - Technical Gap Measurement in a Predictable Process

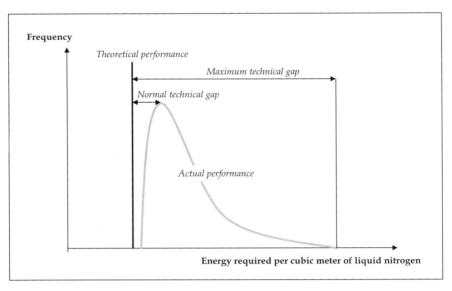

ii. To understand the reason for the shape of the curve and its position, please refer to Chapter 5 on Variation.

150   Manufacturing in Real Time: Managers, Engineers, and an Age of Smart Machines

As an example, this picture may reveal that the normal gap is approximately correct, but that the range of performances is wide and could suggest faults in the compression or separation process. Technical gap measurement will identify not only a potential short-range improvement (narrowing the distribution), but also the longer-term potential of the system (moving the mode to the left).

Many manufacturing processes such as forming, machining, and assembly lead themselves to the same type of calculation. All that is required is a static model of the ideal system identifying CIVs and COVs, and a series of measurements on the actual process performance. Classic industrial engineering measurements of machine times compared to design performance are examples of this process that were introduced by Taylor earlier last century.

In Figure 9.2, the actual performance is a skewed distribution, as would be expected in a process that is controlled close to its capability. Most of the energy measurements are in the proximity of the theoretical performance. It may be possible to narrow further the distribution by process engineering improvements, if the cost is justified.

Figure 9.3 shows another theoretical case of the same process, where the distribution does not conform to the shape expected by a situation under good control. In this case, the measure of technical gap is not meaningful and the practical conclusion is that effort should be spent on determining the special causes of variation rather than trying to improve the gap with theoretical performance.[iii]

---

iii. Even in cases where the process is not in control it is possible to identify qualitatively the potential for improvement by comparing cumulative distribution graphs. This is shown in an industrial example in Chapter 10.

Figure 9.3 - Theoretical Gap Measurement in a Predictable
Process: Another Possibility

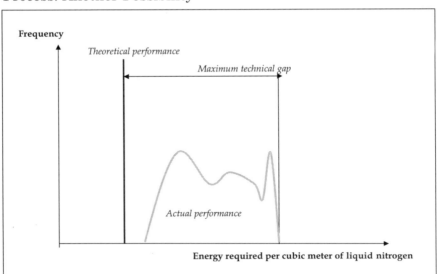

What happens when process definition through physical models is difficult or impossible? In the case of a tire plant, a static physical model is not possible. However, it is possible to build an approximation by the use of dynamic simulation. Dynamic simulation has the capability of accepting ranges of CIVs in either numerical or graphical form, and producing a probable range of COVs. Experience-based guidelines for best practice can be built into a model, and so can any other inputs from the process design or adaptive controllers.

In the case of a dynamic-model reference the technical gap measurement will look like that in Figure 9.4, where we have used the COV of manufacturing cycle time.

Figure 9.4 - Technical Gap Measurement in an Experience-Based Process

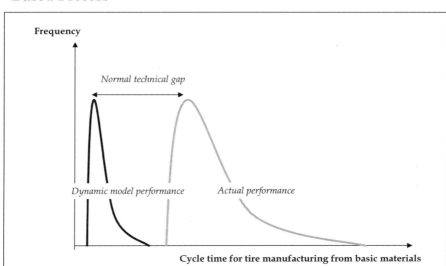

The technical performance gap is now defined by the space between the two distributions. The same reasoning can now be applied to define the improvement potential as in the case of the predictable process.

Understanding the nature of the gap to be closed is a preliminary step to any improvement program or to process and equipment modifications. In the chapter on variation we have defined the causes of variation in two categories:

- Common causes, attributable to the design of the technical process or of the business process

- Special causes, attributable to errors in the operation of the technical process or of the business process.[7]

In the example in Figure 9.2, the shape of the curve suggests that the special causes are under good control and a reduction of the technical gap would require reengineering the process. Technical gap measurement will serve as the base for economic justification of the project and also will clearly identify the project objectives.

In the example in Figure 9.3, measurement of the gap is not informative since the variation in energy requirements is probably due to special causes and the appropriate action would be to undertake a continuous improvement program with the goal of rendering the distribution of results closer to a random walk distribution.

Proceeding with continuous improvement, without knowing the objective for the overall system, is commonly observed practice in the application of Statistical Process Control in manufacturing plants. The same applies to improvements in administrative processes and to functional groups dealing with finance, R&D, IT, and engineering. The functional groups are particularly vulnerable to "local" decision making without a unified strategic vision. This is the result of "local" spatial orientation (manifested in functional chauvinism) and "local" time orientation (manifested in time allocation to tasks with no regard for the organization needs).[8]

It should not be surprising that under these circumstances the results are disappointing. It is equivalent to a group of friends starting off on several random journeys, with the hope of meeting together in Rome in a couple of weeks: unfortunately it is not true that all roads lead to that famous city!

Understanding the gap to be closed will put everybody on the same road map.

Measuring Business Gap

Business gap is an extension of the concept of technical gap, and the principles of its application are the same as those of the experience-based process. The reason is that modeling a business must take into consideration not only quantitative aspects such as production figures, costs, and prices, but also "soft" aspects like consumer preferences and economic trends.

It is in the "soft" aspects that dynamic modeling differs from spreadsheets or hard-wired mathematical formulas. The output of a spreadsheet is a number. This number can be tempered with risk analysis or with confidence limits, but it is still a fixed output. There are many analytical techniques that can be connected to a spreadsheet to improve the visibility of probable outcomes. They include linear programming, utility analysis, decision trees, risk analysis, and more recently, dynamic modeling.

Dynamic modeling has the advantage that inputs can be made variable with time and also expressed as ranges of possible values. As a consequence the output is a statistical distribution of probable results, which can be displayed in simple visual graphics in a time-based framework.

We should also express the concerns of the opponents of dynamic modeling. Dynamic modeling gives the impression of reality transferred into a computer game. This is a potentially misleading interpretation. Computers are digital machines and can only perform calculations on the input they receive. If the problem can be described mathematically, then a dynamic model will assist in calculating solutions (for example, in logistic problems). If the model contains "soft" inputs, the output can become a mathematical expression of opinions. A user knowledgeable of technical or business processes can still be assisted by this kind of model in defining complex interrelationships.

Building a virtual model must be tempered by good judgement. There is presently a wave of disillusionment with dynamic modeling, as is often the case following the initial euphoria with new technologies. This is the result of careless use of dynamic tools to create a strategic plan. The virtual CEO simply cannot exist and possibly will never exist, as computers are limited in recognizing results in complex contexts.[9]

We do not believe that business gap can be measured with the same confidence as technical gap, even using the best tools. The relationships between the many variables are too complex to achieve an exact solution.

A good way to use dynamic modeling is to aim at understanding relationships between different business drivers as an aid to formulating a strategy. To illustrate the point we will chose the example of a new product introduction in an existing market. This model shows where the major issues are in implementation and is helpful for focusing attention on the relationships between risks and rewards. The thinking process, illustrated in Figure 9.5, displays logic similar to Porter's five forces model (see Chapter 2).

Figure 9.5 - The Relationships between Drivers and Outputs in a Dynamic Model

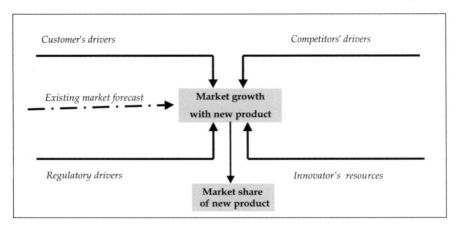

Building a Dynamic Business Model to Improve Decision Making

In this example we will assume that the desired outputs are the overall market growth and the penetration of a new product into this market. The market growth with the existing products is assumed from known projections and is inputted in graphical form for a simulation period of 20 years. The drivers in the model become modifiers of the overall market growth as well as generators of the penetration of the new product.

Selection of the correct drivers is the key to building a helpful model and is an opportunity to share knowledge in a business team. Models built by analysts in a back room are not in this category and only contribute to the skepticism about the new tools. Business modeling should become an opportunity for communication between functional groups and the sharing of a vision; hence the process must evoke a sense of ownership in all contributors to the strategy.

In this respect some high-level modeling tools, derived from the field of biological analysis, are much superior to spreadsheets and industrial engineering simulators because they force transparency of assumptions and model logic. These are the best tools for business gap measurements and can be guided to operate at a conceptual level. On the other hand, strategic software is not suitable for detailed analysis of operations or production systems. We will discuss later different approaches to build the virtual factory.

The selection of the drivers and their quantification is the process of sharing knowledge in the organization, and is an important part of measuring business gap. The drivers in Figure 9.6 were selected by one business as good candidates for analysis. Different businesses may choose others to suit their needs.

Figure 9.6 - Selection of the Drivers Is a Critical Strategic Task

Customer's drivers	Competitors' drivers
• *Quality relative to competition* • *Price* • *Flexibility (customization)* • *Availability (delivery performance)*	• *Size and capacity of existing infrastructure, including supply chain.* • *Brand loyalty* • *Knowledge and capital requirements for new infrastructure*
Regulatory drivers	**Innovator resources drivers**
• *Diffusion of the technology (licensing and antitrust)* • *Regulatory approvals*	•*Skills* •*Money* •*Facilities*

To explain in more detail how the input is done we can choose one of the customer drivers: the quality of the new product relative to existing competition. In this particular case, which dealt with a commodity market, there were very strict and well-established standards and the market was intolerant of deviations. The impact of deviation in quality from that achieved by competition was assessed through discussions with customers. The conclusion was that any deviation in a negative direction would have a serious impact on market share. In contradiction, a quality much superior to that of the competition did not attract significantly increased interest in this particular market segment.

Graphically this situation is represented in Figure 9.7, which was inputted into the model.

Figure 9.7 - Quality Driver of Market Share

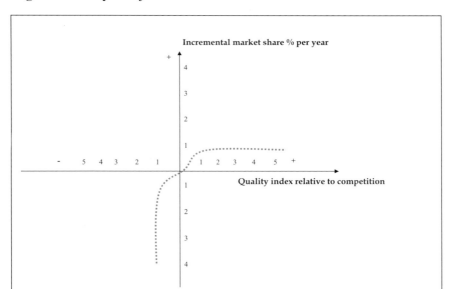

Obviously we are dealing with a "soft" correlation, and this may make the point again of using business modeling results with knowledge and great care. The computer cannot pass any judgment. To a computer any number is a good number! When the input is a combination of 10 or 15 sets of heuristic and numerical data instead of a single graph, it is not possible to see the impact of incorrect assumptions at the output stage.

High-level strategic modeling is a good first step in the design of product streams. It defines the required product attributes from the correct perspective in the supply chain. It also ensures that a business based on a new or improved product stream is a viable business before investments in skills, capital, and other resources begin.

Strategic modeling can be done with spreadsheets and more complex tools. The essential element is the measure of the gaps between the present state of the business and the desired future state. Product streams can then be designed to close the gaps.

The Virtual Factory: Designing Product Streams with Dynamic Models

We have seen that product streams and the flow of resources through them are not constant in time. The amount of variation is a function of the complexity of the stream, and of interactions between streams.

To envisage these interactions it is possible to build virtual models of the stream that include perturbations such as machine failures, interruption of resource supply, product specification changes, and human errors. This can be done in highly automated systems, for either active machine cycles or inactive cycles dealing with logistics, provided there is a continuous measure of the underlying variation.

Models have also been applied to processes that are essentially operated by humans, such as order entry and payroll. The output in these cases is more uncertain.

Dynamic models dealing with complex manufacturing operations are now compact enough to run on a PC. With use and refinement they can become very close to a virtual factory representation of the system. The logic of building a virtual factory is illustrated in Figure 9.8.

Figure 9.8 - Product Stream and Its Virtual Ghost

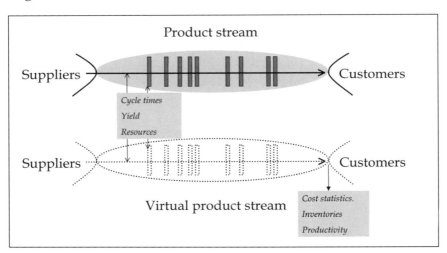

The product stream is first modeled in all its steps of active and nonactive cycles following the methodology described in Chapter 7. The starting point is the product attributes required by the customer. The virtual product stream is then wired to the statistical outputs of the real stream to input cycle times, yield, and resource usage at each step. This wiring can be temporary with the objective of initializing and verifying the model, or it can be permanent. We will first discuss the use of the temporary wiring.

Measuring Gaps Using Virtual Streams

The visible output of a product stream is only the mass flow of goods, which exit the system to the customers. The other ingredients of the product stream (variable cost, capital employed, and information) are not visible without doing some calculations in real time.

Running a virtual product stream is a simple way to do these calculations without "hard wiring" the machines to a complex information system.

This approach is particularly useful if the normal method of measuring performance is based on mass flow (units produced per day and compound yield). Cost calculations in this case can be done occasionally using the virtual stream and can form a basis for standard costing. The virtual stream can also be occasionally wired to customers and suppliers to verify inputs and outputs.

Dynamic modeling allows the measure of technical gaps, even in systems that have a high degree of soft content. This is an excellent application of virtual product streams, since they can be run much faster than real time. For example, zeroing all variations due to failures and running a planned product mix for a year and repeating the simulation several times can measure the ultimate capability of the system. The output of the simulation will be a statistical distribution of variables such as cycle time or mass flow as shown in Figure 9.4. This can be compared to the real stream performance to find how big the "gap to perfection" is.[iv][10] The practical use of this "gap" can be twofold:

- If the gap is small, it is likely that investments in programs such as Total Productive Maintenance or Six Sigma will not be rewarding. Redesigning the stream may be the only way to a step improvement.

- If the gap is sizable or the real stream output is random, it is possible to identify by iteration the parts of the system that have the highest contribution to stream variation and concentrate resources and quality programs on improving these first.

The use of dynamic modeling for product stream design and improvement is a powerful tool and is rapidly gaining acceptance.

Virtual Streams in Real Time

The most advanced use of virtual reality manufacturing is to run the simulated stream in parallel with the real stream and continuously update the simulation with real-life data. A few companies are doing this regularly with great success.[11]

---

iv. In a complex system like a product stream even when all system failures are eliminated there will be fluctuations of cycle times and outputs due to batching, queues, and human process variables.

The basic recommendation of the NGM forum is implemented in this approach, since the simulation provides a tool to explore, in faster-than-real time, the next likely events in the product stream. Rather than guessing the need for intervention in maintenance in material inputs or in process CIVs, the total impact on the system is immediately visible and action can be concentrated on the critical items.

The virtual reality factory will become a major tool in sharing a unified vision of manufacturing, linking people, technology, and business processes in a readily accessible display of all the events taking place in a stream.

It will also open the opportunity for imaginative approaches which could only be tried at great risk on the real system. Experimenting on a real stream may cost millions of dollars and only functional experts are allowed to do it. Experimenting with the virtual factory has no impact on customers. It will bring people together to develop common solutions and improvement goals.

One day the virtual factory may be the only interface of man and machine.

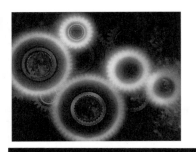

Chapter 10

Industrial Engineers and
Intelligent Machines

Sizing the Gaps

A heat-exchanger manufacturer produces preshaped refrigerant tubes for a number of vehicles. Some of the tubes for the larger sport utility vehicles (SUVs) may need to be over 3 meters long to take refrigerant to the heat exchanger in the rear passenger compartment. The shape is very complex as it must fit around the vehicle unibody frame.

The process starts from coiled aluminum tube, which is cut to the required size, straightened, then processed through a number of machines that swage or braze quick-disconnect ends to the tubes. Finally the tubes go through a numerically controlled bender to give them the desired shape. They are then packaged in tailor-made, returnable plastic containers.

The particular company, which we will refer to as TubeX for the duration of this discussion, produces about 40 products of different dimensions and shapes. The production process is labor-intensive, in spite of the high degree of automation.

This is a case where automation is used to achieve the required degree of accuracy in a technically demanding application involving numerous products. Numerically controlled machines lead themselves to quick product changes and agile manufacturing; they also require highly trained personnel to reset the process for different products and to ensure that the critical output variables of the process remain within the control limits.

Component manufacturing of this kind is an example of agile manufacturing, with strong supply chain links to the car assemblers, integrated quality, and environmental systems (QS9000 and ISO14000), and good statistical process control procedures.

Everything regarding the operations of this company appears to be close to textbook methodology, with the exception of their financial performance. Product prices in component manufacturing are continuously under downward pressure, because these businesses are part of a supply system that is global and highly competitive.

Effective manufacturing does not necessarily mean profitable manufacturing, and the case of TubeX is a good one to examine.

As shown in Chapter 6, profitable manufacturing is the result of the statistical distribution of costs and prices. It is possible that many component manufactures find themselves in the position shown in Figure 10.1: profitable some of the time, but not enough of the time.

Figure 10.1 - TubeX Profitability Is Driven by Performance Distribution in Cost and Price

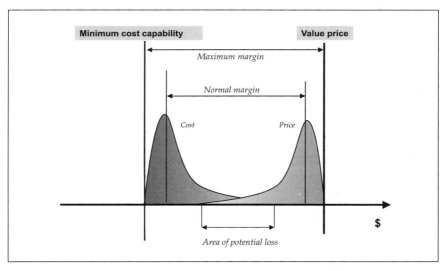

TubeX has instituted process control procedures. The statistical process control charts are not in real time. As a consequence TubeX can tell if a process is drifting from its targets over a period of time, but cannot measure instantaneously the gap between present performance and the maximum capability of the machine systems. Most of the costs within the manufacturing operation of TubeX are fixed (labor included, if the number of shifts is constant) so that cost is a function of material requirement and the cycle time for each component passing through the line. Since real-time cycles are not known, TubeX decided to establish the ultimate system capability in component cost with the objective of making better-informed strategic decisions.

TubeX designed specific product streams and modeled these streams dynamically to establish the gap between existing performance and the ultimate potential of the company. The process involved the following:

- Product stream design for each product
- Scheduling sequences
- Real-time measure of machine or process failures to perform required quality specs
- Statistics of setup times and machine breakdowns
- Individual product yield
- Dynamic modeling

Figure 10.2 represents a modeled 4-week production period based on the statistical performance of the actual plant and shows the output of the dynamic model in terms of the cycle times of a single product (source: CAMM).

Figure 10.2 - Cycle Time in Plant Production and an Optimized Model

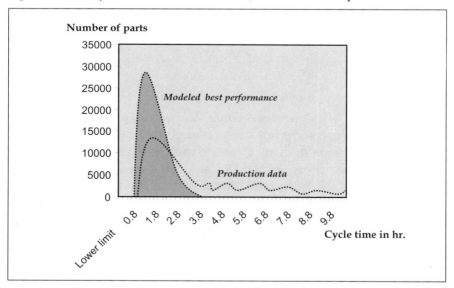

Figure 10.2 also shows the best performance possible with this particular manufacturing system. The optimized model is a

"random walk" distribution with a maximum cycle time of about 3.8 hours. The production data are random and have cycle times as high as 11.0 hours. This particular curve shape indicates a chaotic behavior due to failures, and it is difficult to measure the gap to the ultimate performance. A better way to achieve this is to plot the cumulative distribution in the two cases, and measure the area between the production data and the optimized curve, as shown in Figure 10.3.

Figure 10.3 - Cumulative Distribution of Cycle Times

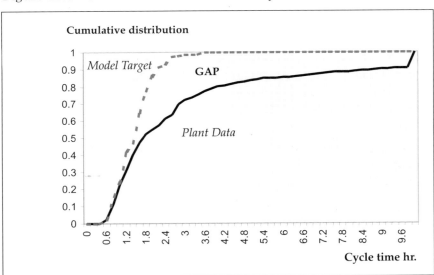

The gap measuring process, applied to several products, allows TubeX to direct its training and engineering efforts toward the product streams, which show the largest improvement potential in variation. Without this vision the company might have been tempted to invest more capital to improve productivity.

The situation of TubeX is an example that illustrates how the statistical performance of the system in real time is a driver of productivity and cost. This kind of analysis was not possible only a few years ago, since both data collection and dynamic models required an inaccessible level of computation. If Shewhart and Deming had the computing tools at their disposal that we have today, the methodology of statistical process control would have been designed in real time rather than on historical data.

In addition to their basic function of producing precise products, sophisticated machines can record their own performance continuously, without major investments in equipment. There is almost no manufacturing machine that does not have a programmable logic controller (PLC), and any PLC can be connected with a laptop to process real-time measurements. The only difference between complex and expensive machines (such as a half-billion-dollar steel mill) and the simple swaging press is that this kind of analysis could be done continuously on the more expensive machine, as opposed to occasionally on the other. Great advances have been made in automation and networks, and yet most operations are run as they were before information networks were created, with good process control, good team organizations, and good historical records. This approach does not lead to the ultimate potential of a manufacturing system.

The manufacturing industry is far from adopting real-time measurements, and this is one of the reasons why the productivity of capital assets in manufacturing is decreasing.[i] This transition is difficult because:

- Quality systems in manufacturing are often adopted for compliance with customer requirements

- Flow in discrete processes is not an intuitive concept; it requires training

The Pressure to Comply

An unusual sight, when driving through a modern industrial park, would be a manufacturing plant with no banner proclaiming achievement of one of the many industrial quality standards. One

---

i. The few examples in advanced companies are a beginning, but not yet the norm. Their results are not visible in overall manufacturing productivity. The predictions and recommendations of the Next Generation Manufacturing Forum of 1999 will probably take another 10 years to materialize.

banner may state, "We are proud of our XXXX Quality award," another "We are a XXX company." A third one may claim a list of "Supplier of the year awards." Who is the intended recipient of this information? It could be the employees, visiting customers, potential customers, or even the general public.

There is little doubt that compliance with certain quality standards has an impact on customers; that is, and has always been, the prime motivator of quality systems. They provide the customer with a reassurance that the company is capable of adopting the best practices for product consistency and rapid response to possible problems. The system also becomes a language through which the company can communicate with customers and suppliers to make the successful operation of supply chains achievable. Quality systems that look at the manufacturing business as a total entity are excellent tools through which to see the flow of products and to demonstrate it to customers and suppliers.[1] In the process, there is also an understanding of shared values, including financial rewards. For companies who make the commitment to quality systems (and it is a large commitment), moving to real-time control and machine-generated information is a natural evolution.

The majority of manufacturing today is not in this league: the quality banner is simply a cost of doing business and complying with the requirements of sophisticated customers. Even the well-intentioned manufacturing operations, which look at machines and plants as self-standing units and ignore the softer issues of management and administrative practices, will not achieve much more than good marks for compliance. Profit will remain an elusive goal.

Lean manufacturing has come under criticism because it does not generate profits. The conclusion has to be that it does not, unless it is applied to the company as a whole and implemented with relentless discipline. Japanese companies such as Toyota, which pioneered this approach, have gained a worldwide reputation for product value and as a consequence have gained market share. Toyota has gone from a local Japanese company to number 11 among global corporations in 2001 (an impressive performance well documented as a result of innovative manufacturing practices). In the early years, Toyota did not perform as well for its investors, returning meager profits as a result of concentrating exclusively on manufacturing and maintaining a heavy company superstructure at the same time. All this has changed in the past few years, as Japanese business methods and financial performance measurements have aligned with Western practices in the global economy. Toyota has responded by applying its knowledge to the overall corporate structure.[ii]

As to the use of banners as a way to motivate the workforce, recent research suggests that this may not be a very long-lived incentive. The pride of achievement and the novelty wears off as quickly with the workforce as it does with the general public.

Customers and investors value quality systems. Some companies may not need or be able to justify any broader application. To go past this point involves the high degree of discipline and commitment we have described in this book. Anything less will not be reflected in the company bottom line.

There is a reward to managing a manufacturing business as a single system, not only in physical assets, but also increasingly in the intangible values, which are created by knowledge. Knowledge means educated employees, which in turn means innovation from product research and strategic positioning based on understanding in depth competitive threats and opportunities.

> "Successful manufacturing will evolve into system management and move away from process management as the investment community realizes the value of intangible assets."[2]

---

ii. The 2000 Toyota Annual Report showed that in 2000, Toyota had a net profit of 3.1% of sales of $121,334 billion USD, compared with Ford Motor Company net profit of 2% and GM Corporation at 2.3% of sales of approximately $170 and $180 billion USD, respectively.

Man, Machines, and Product Flow

We are biological beings and flow is a continuous process in our bodies that must take place with no pause, or we will die. We are not naturally inclined to perceive this flow. We are aware of the here and now in the form of perceptions and mental images and only specialized training can change this.[3] Think of playing music, understanding navigation, or studying history as examples of how we can be trained to connect events into flow. Some talented people, such as race-car drivers or fighter-jet pilots, are capable of linking instantaneous events with flow.[iii] For most of us an understanding of flow in quantitative terms is not achievable without the help of a machine.

Machines are at the opposite end of the spectrum; they cannot perceive the here and now in a contextual sense, but are eminently adapted to describing flow in quantitative terms, even in extremely complex systems. This is the inherent reason why the future of manufacturing will likely be driven by machines, which can communicate with people easily and intelligently.

A 1994 article on the factory workplace displays an interesting statement about men and machines: "Increasingly, smart humans are replacing dumb robots."[4] The enthusiasm of teamwork and the results it can create were the great cry of the time. However, intervening years have proven the opposite to this prophetic statement. People do not want to replace robots. They want the robots to help them to do repetitive tasks in a smart way, allowing humans to concentrate on creative activities.

In the chapter on automation we described a functional diagram of hierarchical automation systems as originally described by Allen-Bradley, a well-known supplier of automation and control systems. It is a useful way to understand the flow of information but in the

---

iii. The transfer of information in a rally car from navigator to driver and the transmission of that information to driving performance seems seamless. The fact the car is traveling faster than the information can be processed and input is what makes the talent of the driver remarkable.

past few years it has become increasingly oversimplified. Machines are evolving faster than we thought in the direction of distributed systems, where many complex functions can be performed locally and the connectivity of the enterprise is achieved through network communications.

The largest impediment to the implementation of connected enterprises is still the lack of standard communication architectures and open systems.[5] Rapid progress is being made in this area. Centralized systems, whether in finance, Enterprise Resource Planning (ERP), or factory process control, are rapidly being replaced by "granular" architectures in which the local activity is tailored to the needs of a single operation and the network takes care of the system integration. This gets closer to the working of organic systems, and the prediction of Allen-Bradley expert Dr. Odo Struger is that people and machines will all participate in these "holonic" manufacturing systems[iv] as an integral part of far flung networks.[6]

> We are learning to see flow by talking to the machines that make it happen.

The Global Manufacturing Business and the Measures of Productivity

Productivity trends in North America, for all businesses excluding agriculture, have shown great promise in the last 5 years of the millennium, primarily as result of the impact of information technology on Gross Domestic Product (GDP). Economists are debating whether this trend shows a permanent sign of sustainability or is a result of a prolonged "boom" economy. This book is

---

iv. "A holon is an autonomous, intelligent, distributed, cooperative device. It could be a robot, a motor, a person, or any other component of a process. Holonic systems are modeled after biological organisms and display such features as intelligence, autonomy and cooperation."[7]

not about the here and now, but about the important permanent changes that the information technology miracle has brought about in manufacturing.

Figure 10.4 shows the annual changes in productivity from 1965 onward. Even accounting for a possible boom effect, the results suggest that the new e-society may allow for a steady growth of 2-2.5% per year, which is much better than the 1975-1995 period when the information revolution was just taking place.

Figure 10.4 - Yearly Percent Change in Productivity

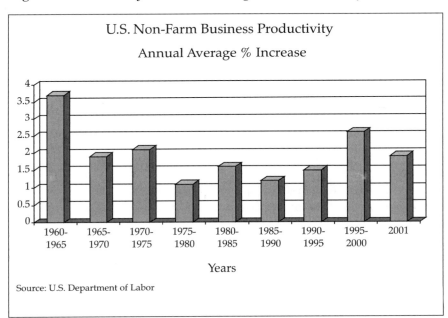

More disturbing is the fact that several industrial sectors show a decline in the productivity of capital employed (measured as $sales/$capital employed). The U.S. Federal Reserve has concluded in recent studies that almost half of the productivity improvements in the 1995 - 2000 period were driven by "capital deepening" (for example, substitution of capital for labor). If this conclusion is correct, a measure of Total Factor Productivity (TFP), which takes into account capital and labor, would show a decline from 1970 to 2000. The decline in productivity changes in 2001 is largely due to the lack of capital investment in a recessionary economy: further confirmation of capital dependence.

There is a substantial body of research trying to explain these trends, which focuses on two factors:[8]

- Productivity figures, measured by traditional statistics and accounting, are questionable in the new economy

- Business is in transition, trying to capture the value of new computer and network technology

In the case of manufacturing (which is faring better than service businesses in real productivity improvements), the results of heavy investments in automation and networks have only now started to be visible. This should not be a surprise since the implementation cycle of completely new technology, like electrification in industry in the 1920s, can take 20 years.

It could also be true that the measurements developed by the early industrial engineers are no longer valid. In school, engineers are taught the discipline of measuring using the same yardstick, no matter what the circumstances, or else they end up with the Tower of Babel! But engineers rarely measure "soft" things like knowledge and information. Measurement of soft issues requires a degree of flexibility.

> Manufacturing industry has made massive investments in computer technology for the past 20 years. Productivity is likely to increase rapidly not only in tangible goods but also in soft values, which we are ill-prepared to measure, such as knowledge, technology, and skill assets.

If we look at the stock market in the past few years the ratio between market value and net asset value was greater than six for the S&P 500 selected companies. This means that the stock market values intangible assets much more than the tangible assets measured by conventional accounting procedures. Economists estimate that in 1929 the ratio of intangible assets to tangible assets was 30% to 70%. By 1990 it had risen to 63% to 37% and is still rising rapidly.[9] This number suggests that the math we are using to measure productivity may be wrong in two ways:

- The real GDP adjusted for the creation of soft value could be much greater.
- The assets employed to generate the GDP could also be much greater.

There is no evidence to suggest that these statements apply equally in all industries. Manufacturing has already spent huge monies to create soft assets; it now needs to extract the value from the investment.

People, Machines, and the Workplace

Lynda is going for a hayride during her lunch break. The company she works for has organized the hayride with large Clydesdale horses and a beautifully decorated wagon full of clean hay. The wagon is not going very far, winding its leisurely way around the research camp, between lily ponds and pleasant tree groves. Lynda can relax and chat to her fellow employees before returning to her computer and another grueling afternoon of writing complex software. The project is due a week from now in quality control, and then to an anxious customer. It is not likely that Lynda will have much more relaxation within the next 10 days, unless she leaves her very well remunerated job, or the market for logistics software suddenly takes a plunge. Lynda is a Highly Qualified Professional (HQP) and she is a hot commodity in an economy in search of rare skills. The hayride is only a small token of what her employer is prepared to do to keep her satisfied and productive. It is also a symbol of a flexible and skill-oriented organization taking big risks and hoping for big rewards.[10]

Bud is sitting outside the shipping door of the casting plant on a picnic table provided by the company and is eating his home-prepared sandwich. The sun is beating mercilessly on the dusty gravel and Bud is trying to relax during his 20-minute paid lunch break. He can extend the lunch break another 20 minutes without pay, and is contemplating doing so. He decides that $7 is too high a price to pay, when he can go home in a few hours and relax under a tree. Finishing his sandwich, Bud returns to his post at the forging press. Bud is a member of a production team with good experience and training that is an important asset to his company, and enjoys regular working hours with occasional overtime. Through good and bad times the plant will try to keep Bud on its payroll because he represents an important part of the knowledge necessary to run the business. This business is a steady, low-risk, and low-rewards operation that has kept up with technology and organizational changes, thriving from improvements in productivity and cost. It is typical of many part suppliers and durable goods manufacturers.

These two examples are not for the purpose of comparing lifestyles: we want to point out the common values of these two individuals. Both perform essential roles for their business, both have a unique skill to offer, and both embody a part of the company's intangible value.

These two employees are typical of 60% of today's workforce, and the shortage of the skills they represent is the most serious problem that North American manufacturing is likely to encounter in the next decade.[11]

> "Since World War II, there has been a quest in America for everyone to have a college degree, and a great deal of money has been put toward that goal."[12]

The result is that college education has increased, but a university degree is not necessarily the type of education required in 60% of the jobs. Manufacturing industry requires a mix of skills ranging from Highly Qualified Professionals to skilled operating technicians. The HQPs may represent 10% or less of the need, even in highly technological operations. Individuals like Bud, trained either on the job or with a technical college or apprentice school degree, will be the core of the future workforce. The need will gradually increase in the next decade and unskilled jobs will probably represent less than 10% of manufacturing.

> The future of productivity improvements based on the computer and automation infrastructure, which is already in place, is totally dependent on the technical skills that manufacturing industry can deploy and retain. The greatest need is for educated technologists and not for university graduates.

We are in an important transition period, which is poorly understood by educators and the young people planning to join the workforce. Unskilled jobs are disappearing rapidly and are being replaced by smart machines. Even mechanical floor sweepers can now navigate a plant on their own. It is not a question of labor cost: the manufacturing industry cannot afford the errors produced by unskilled employees, nor can it afford to invest a disproportionate amount of energy in training. Europe, Japan, and South America have developed effective ways to produce the right kind of skilled employee without university degrees. North America needs to do the same.

The lifestyle of the employee is also dependent on his/her education. A trained technologist has the same job mobility as an engineer, but will probably choose to remain in the same community and be part of a local pool of skills. He/she will also derive more satisfaction than a college graduate from the tasks required in lean/agile manufacturing, which deal with technological details and with communication with machines. The new management structures will also ensure that technologists will have a say in the quality of life on the job. These are the jobs of the future and are rapidly replacing the old manufacturing hierarchy.

Manufacturing also needs engineers, accountants, and MBAs well-trained in global thinking and motivated by innovation and improvement. They represent another intangible asset in addition to an educated operating workforce. Manufacturing does not need many of them, but needs good ones, even if they are expensive.

Lynda and Bud are part of this future. Lynda will master the use of machines to make her life manageable and Bud will probably get a tree and a nice lawn for his lunch. Both of them are looking at a brighter future.

The Preextinction Society

In 1798 Thomas Malthus published his essay *On the Principle of Population*, which described his "population principle": "the constant tendency of all animated life to increase beyond the nourishment provided for it."[13] Though Malthus later mitigated this statement with a number of natural and moral restraints, his work left a lasting impression of a humanity condemned to die from lack of resources. Malthus was a scientist and an accomplished statistician. His work was based on the information available at the time on the present and potential productivity of agriculture and industry. He predicted an almost geometrical increase in population, which starvation would contain, and only a linear increase of resources.

Industry in the early nineteenth century was operating at approximately 10% efficiency in energy and resources. Agricultural practices were similarly wasteful in depleting usable land and pasture; thus it is not surprising that Malthus used a linear increase of goods needed for sustenance. He could not have envisaged the capability of supporting 6 billion people on the planet.

With some provisos, the question of balance between population and resources remains unchanged and is reshaping the attitude of industry and of society. This attitude was highly visible in the Hanover Expo 2000, a carefully staged exhibition with a conservation theme warning of the risks of squandering limited resources and the potential for extinction. Manufacturing and the methods used to manufacture and distribute goods are at the core of using our resources more effectively and maintain an indefinite balance in Malthus's equation. If we are successful, our society can prevent further extinction of other species and still provide adequate sustenance for mankind for a long time.

The potential is there. An extensive database produced by Delft University in Holland[14] shows that many materials are still produced with poor energy and material efficiencies, in comparison to what is theoretically possible (a ratio of one is the theoretical minimum and any higher number represents proportionately more energy). The gap is in some cases so large that the same materials and energy could produce many times the amount of goods produced today. Figure 10.5 shows the excess work contained in several materials against thermodynamic requirement. The Japanese call this Exergy. The chart illustrates a potential and does not suggest that all materials can be made with higher efficiency at an acceptable cost.

Figure 10.5 - Exergy Ratio in Different Materials, Wood Being Closest to Natural State

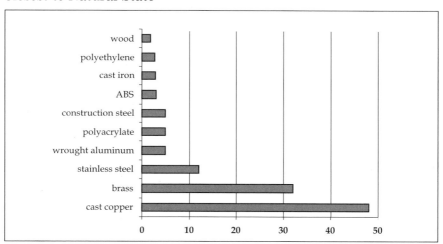

In the long run, accessing the opportunity of greater efficiency not only makes ecological sense, but also makes good business practice. Businesses adopting a "green" approach are able to produce quality goods with a high level of profitability. An excellent example are the steel mini-mills, which have consistently outperformed the integrated manufacturers in profit and in service for the past 15 years. The mini-mills are new investments and have a disadvantage against the existing infrastructure in the cost of capital assets, but their efficiency is more than offsetting the initial investment.

> The next generation of global business is shifting its focus to increased efficiencies in the use of resources and, in this process, is becoming the key to sustainability of the natural environment and of quality of life.

The principles of "green" manufacturing, which determine its long-term potential, are tied to the supply chain concept and to the capability of global communication. The following are its most important traits:

- Design products on the basis of life-cycle analysis[v]
- Design supply chains for minimum cycle time
- Operate and control manufacturing in real time (product streams)
- Design for 100% reuse of all materials

The importance of flow is apparent in supply chain and manufacturing operations, but it is also the foundation of material recycling and reuse. Fast flow is also energy efficient flow, as we have seen in the chapter discussing product streams, and leads to a minimum of waste. Effective recycling can only be achieved by designing recovery of the material and energy content of the product into a predesigned flow.

This vision may appear inconsistent with the picture of depredation and abuse of land, forest, and water systems perpetrated by global corporations, according to some environmentalist news. Companies that behave irresponsibly do exist, but they will not exist for long: waste simply does not make economic sense. Meanwhile the majority of the global corporations have found a new business in prolonging the use of scarce resources and finding new ones. Fuel cells, solar and wind power, and composite materials made from recyclable or renewable sources are examples of this.

---

v. Life-cycle analysis is an impact study over the useful life of a product from the procurement of the raw materials, through its use and finally its disposition.

> The opportunity to steer manufacturing toward a sustainable balance is an exciting and profitable prospect.

Dealing with Globality

"Unless we can come to terms with the global-image economy and the way it permeates the things we make and see, we are doomed to a life of decorating and redecorating," writes Bruce Mao, one of the great designers and social students of our times.[15] Mao wants us to accept globality as the inevitable direction that humanity decided to take 40 years ago, and as an irreversible direction. Whether this is good for the local business or for local culture depends on our ability to accept globality and adapt to it, rather than passively responding to every new trend.

In manufacturing, the global market takes care of competitiveness regardless of local interests. An uncompetitive enterprise supports waste, lack of vision, and inefficiency. Ultimately nothing could be more damaging to the use of resources and to the environment than an uncompetitive plant or retail operation.

The methods we have learned in the past 50 years, now supported by smart machines, are opening the opportunity to provide goods to an increased population using fewer and fewer resources. We have also learned to measure how well we can do and the gaps we have to close. We have no other alternative. Craft manufacturing, organic farming, and "simple" lifestyles are too wasteful to contemplate them on a large scale. These are the lifestyles that induced Malthus to conclude that a large part of humanity was doomed within a few centuries. Too many things have changed and there is no turning back to nature. We may find that "nature" is no longer there or that it has changed so significantly as to be totally hostile to our way of life.

We do not need to mine asteroids: all of our materials are reusable. Energy in the solar system will last for billions of years and we are learning to capture it. The World Wide Web is bringing together the thinking of 1 billion people. We have had these opportunities at no previous time in history.

The New Generation of Industrial Engineers

This book is about manufacturing useful products in an efficient and profitable way, through an advanced use of smart machines.

The goal of efficiency, including limiting the use of scarce resources, is compatible with providing value to customers and at the same time making an adequate manufacturing profit. This is demonstrated in the dynamic behavior of manufacturing systems in the form of "gaps to best practice" and in the process of closing these gaps.

Using competitors in the same industry as a benchmark to best practice is only a partial step in appreciating the scope of improvement opportunities. Modeling and computer simulation offer better tools to assess the ultimate degree of perfection that a business can attain.

There is a new generation of industrial engineers, trained in supply chain systems and in the use of sophisticated analytical tools. They are the future designers of business processes and their effectiveness is dependent on their understanding of the strategic goals of the organization.

In this book we have used the design of product streams as a key to communicate effectively at all levels of an organization, from the engineers who talk to the machines to the leaders of the company strategy. If the concept appears too simple, we will have succeeded in conveying the basic idea that manufacturing business processes are essentially simple, but embody some of the most complex technology that humans ever invented. Product streams and their dynamic interpretation in statistical models create a language bridge between technology and business, and a communication tool within organizations.

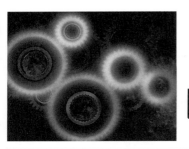

References

Chapter 1

1. "Manufacturing." <u>Encyclopedia Britannica Online.</u> May 29, 2000.
   <http://search.eb.com/bol/topic?eu=51905&sctn=1>

2. "Mass Production: Mass Production and Society." <u>Encyclopedia Britannica Online</u>. June 15, 2000.
   <http://search.eb.com/bol/topic?eu=109274&sctn=8>

3. "Mass Production: Mass Production and Society."

4. Dihel, Eidth. <u>Bookbinding: Its Background and Technique</u>. Vol. II. New York: Dover Publications Ltd., 1980. P. 104.

5. Rybczynski, Witold. <u>One Good Turn: A Natural History of the Screwdriver and the Screw</u>. Toronto: HarperCollins, 2000. P. 75.

6. "Mass Production: The Industrial Revolution and Early Developments." <u>Encyclopedia Britannica Online</u>. June 15, 2000.
   <http://search.eb.com/bol/topic?eu=109274&sctn=2>

7. "Industrial Revolution." <u>Encyclopedia Britannica Online</u>. June 15, 2000.
   <http://search.eb.com/bol/topic?eu=43323&sctn=1>

8. "Child Labor." <u>Encyclopedia Britannica Online</u>. June 15, 2000.
   <http://search.eb.com/bol/topic?eu=24432&sctn=1>

9. Hammerton, J.A., ed. <u>Harmsworth's Universal Encyclopedia.</u> London: The Educational Book Co. Ltd. P. 4233.

10. "Industrial Revolution."

11. "Mass Production". <u>Encyclopedia Britannica Online</u>. June 15, 2000.
    <http://search.eb.com/bol/topic?eu=109274&sctn=1>

12. "Mass Production: The Industrial Revolution and Early Developments."

13. Cahill, Marie. <u>A History of Ford.</u> New York: Smithmark Publishers, 1992. P. 36.

14. Cahill. P. 21.

15. "Ford Focus sets a new standard for small cars at a great low price." <u>Autoworld.</u> National Automobile Bankers Association/Vehicle Information Services, 1999. August 15, 2000.
    <http://autoworld.com/news/Ford/Focus_Price.htm>

16. The Boyd Company. <u>Metalworking Industry Study: "A Comparative Cost Analysis for Metalworking Operations."</u> New Jersey: The Boyd Company Inc. P. 39.

17. Moody, Patricia E. <u>Leading Manufacturing Excellence</u>. USA: John Wiley & Sons Inc., 1997. P. 26.

18. Moody. P. 26.

19. Toyoda, Eiji. Toyota: Fifty Years in Motion. New York: Kodansha International, 1987. P. 65.

20. Toyoda. P. 65.

21. Womack, James P., Jones, Daniel T., and Roos, Daniel. The Machine that Changed the World. Ontario: Collier Macmillan Canada Inc., 1990. P. 48.

22. Kilian, Cecelia S. The World of W. Edwards Deming. 2nd Ed. Tennessee: SPC Press, Inc., 1992. Pp. 1-12, 29.

23. Kilian. P. 10.

24. "Automation: Robots in Manufacturing." Encyclopedia Britannica Online. June 16, 2000. <http://search.eb.com/bol/topic?eu=117180&sctn=14>

25. "Automation: Robots in Manufacturing."

26. Moody. P. 59.

27. "Artificial Intelligence." Encyclopedia Britannica Online. June 16, 2000. <http://search.eb.com/bol/topic?eu=9829&sctn=1>

28. Warwick, Kevin. In the Mind of the Machine: The Breakthrough of Artificial Intelligence. Great Britain: Arrow Books Ltd., 1998.

29. Karlsson, Reine. Life Cycle Considerations in Sustainable Business Development: Eco-Efficiency Studies in Swedish Industry. Goteborg: Chalmers University of Technology, 1998. Paper V, P.1.

Chapter 2

1. "Integrated Supply Chain Management Program." Massachusetts Institute of Technology. June 18, 2001. <http://web.mit.edu/supplychain/www/ouiscm/base.html>

2. Friedman, Thomas L. The Lexus and the Olive Tree. New York: Anchor Books, 2000. Pp. 93-95.

3. Porter, Michael E. Competitive Strategy: Techniques for Analyzing Industries & Competitors. New York: Free Press, 1990.

4. Hammonds, Keith H. "Michael Porter's Big Ideas." Fast Company. Issue 44. P. 150.

5. Hays, Constance L. "Coke Tests Vending Unit That Can Hike Prices in Hot Weather." June 21, 2001. <http://www.philipkdick.com/articles/102899coke-vending.html>

Chapter 3

1. Goldratt, Eliyahu. The Goal: A Process for Ongoing Improvement. USA: North River Press, 1984.

2. Lepore, Domenico, and Cohen, Oded. Deming and Goldratt: The Theory of Constraints and the System of Profound Knoweldge. USA: North River Press, 1999. Pp. 106-111.

3. Goldratt, Eliyahu. Throughput Accounting. USA: North River Press, 1998. P. 32.

4. von Beck, Ulrich, and Nowak, John W. "The Merger of Discrete Event Simulation with Acitivity Based Costing for Cost Estimation in Manufacturing Environments." Department of Mechanical and Industrial Engineering. University of Illinois Urbana-Champaign. Simulation Conference. Winter 2000.

5. Stewart, G. Bennett III. P. 3.

6. Cahill, Marie. A History of Ford. New York: Smithnark Publishers, 1992. P. 21.

Chapter 4

1. Moody, Patricia E. Leading Manufacturing Excellence. USA: John Wiley & Sons, Inc., 1997. P. 34.

2. Rusk, James. "Fido's Lost Fees Cost City $700,000." The Globe and Mail. February 15, 2001. P. A1.

3. "Pygmalion and Galatea in Myth." Mythography. June 15, 2001. <http://www.loggia.com/myth/galatea.html>

4. "Golems." Compton's Encyclopedia Online. June 15, 2001. <http://www.comptons.com/ceo99-cgi/article?'fastweb?getdoc+viewcomptons+A+17915+0++Golem'>

5. Warwick, Kevin. In the Mind of the Machine. Great Britain: Cox & Wyman Ltd., 1998. P. 43

6. Kaufman, Debra. "The Roboteer." Wired News. June 15, 2001. <http://www.wired.com/wired/archive//6.06/eword.html?person=karel_capek&topic_set=wiredpeople>

7. Johnsen, Edwin G., and Corliss, William R. "Teleoperators and Human Augmentation." Gray, Chris H., Figuroa-Sarriera, Heidi J., and Mentor, Steven. The Cyborg Handbook. Great Britain: Routledge, 1995. P. 82.

8. Johnsen, Edwin G., and Corliss, William R. P. 85.

9. Sobel, Dava. Galileo's Daughter: A Drama of Science, Fath, and Love. Fourth Estate Ltd., 1999. P. 27.

10. "Abacus to Eniac: Highlights in the History of Computing." <u>University of Pennsylvania.</u> June 15, 2001.
<www.upenn.edu/computing/printout/archive/v12/4/pdf/abacus.pdf>

11. Haugeland, John. <u>Artificial Intelligence: The Very Idea</u>. Massachusetts: MIT Press, 1985. P. 53.

12. Dawkins, Richard. <u>The Selfish Gene</u>. Oxford: Oxford University Press, 1989. P. 23.

13. Warwick. P. 95.

14. Haugeland. P. 170.

Chapter 5

1. Harry, Mikel and Schroeder, Richard. <u>Six Sigma: The Breakthrough Management Strategy Revolutionizing the World's Top Corporations</u>. New York: Random House Inc., 2000. P. 39.

2. Wheeler, Donald J. <u>Understanding Variation: The Key to Managing Chaos</u>. Tennessee: SPC Press Inc., 1993. P.4.

3. <u>Lucent Technologies</u>. September 25, 2000.
<http://www.lucent.com/work/family/docs.shewhart.htm>

4. Gould, Stephen Jay. <u>Full House: The Spread of Excellence from Plato to Darwin</u>. New York: Three Rivers Press, 1996. P. 150.

5. <u>Lucent Technologies</u>.

6. <u>Lucent Technologies</u>.

7. Besterfield, Dale H. <u>Quality Control</u>. 5th Ed. New Jersey: Simon & Schuster, 1998.

Chapter 6

1. Corbett, Thomas. <u>Throughput Accounting</u>. USA: North River Press, 1998.

2. Gross, Irwin. Lecture notes from the Senior Marketing Management Program through the Alcan Employee Education Center, given April 12-15, 1999, Detroit Michigan.

3. "Value Pricing Pilot Project." <u>U.S. Department of Transportation - Federal Highway Administration.</u> December 8, 2000.
<http://www.fhwa.dot.gov/policy/vppp.htm>

4. "Value Pricing Pilot Project."

5. Moody, Patricia E. <u>Leading Manufacturing Excellence.</u> USA: John Wiley & Sons, Inc., 1997. P. 38.

6. Harry, Mikel, and Schroeder, Richard. <u>Six Sigma</u>. New York: Random House, Inc., 2000. P. 150.

Chapter 7

1. Schonberger, Richard J. <u>Building a Chain of Customers</u>. Free Press, 1990.

Chapter 8

1. "Chapter 4 : Project Organization." <u>Southern Polytechnic State University</u>. March 27, 2001. http://www2.spsu.edu/tmgt/richardson/ProjectManagament/Lectures/Chap-04/Chap-04.html.>

2. Thomas, Lewis. <u>The Fragile Species</u>. New York: MacMillan Publishing Co., 1992.

3. Rivard, Susan. Opening comments for the 3rd International Engineering Conference "Integration of Human Resources and Technologies: The Challenge." Montreal, Quebec: May 1999.

4. Rivard, Susan.

5. Womack, James P., and Jones, Daniel T. "From Lean Production to Lean Enterprise." *Harvard Business Review: On Managing the Value Chain.* USA: Harvard Business School Press, 2000. Pp. 225-226.

6. Womack, James P., Jones, Daniel T., and Roos, Daniel. *The Machine That Changed the World.* Ontario: Collier Macmillan Canada, Inc., 1990.

7. Bouzon, Arlette. "De l'emergence de l'espertise et de l'innovation dans let organisations." Proceedings from the 3rd International Industrial Engineering Conference - "Integration of Human Resources and Technologies: the Challenge." Montreal, Quebec, May 1999. Pp. 461-470.

8. Bouzon, Arlette.

9. Schenk, Christopher, and Anderson, John. <u>Reshaping Work 2</u>. Canada: Canadian Centre for Policy Alternatives, 1999. P. 87.

Chapter 9

1. "Next-Generation Manufacturing: A Framework for Action." *Executive Overview.* January 1997. August 15, 2000. <http://www.dp.doe.gov/dp_web/documents/ngm.pdf>

2. "Next-Generation Manufacturing: A Framework for Action." P. 5.

3. Aeppel, Timothy. "Mounting Pressure under Glare of Recall, Tire Makers Are Giving New Technology a Spin." The Wall Street Journal. March 23, 2001. P. A1.

4. Aeppel.

5. Annual Report 1999: Air Liquide. P. 7.

6. Phillips, Mark. "The Incredible Tire Journey." American Rubber Technologies, Inc. April 10, 2001.
   <http://www.americanrubber.com/recycle_today_article.html>

7. Besterfield, Dale H. Quality Control. 5th Ed. New Jersey: Simon & Schuster, 1998.

8. High Performance Systems Inc. Getting Started with the ithink Software: A Hands-On Experience. High Performance Systems Inc., 1996.

9. Haugeland, John. Artificial Intelligence: The Very Idea. Massachusetts: MIT Press, 1985. P. 213.

10. Keenan, Gregory. "Six Degrees of Perfection." The Globe and Mail. December 20, 2000. P. M1.

11. "APACS". PRECARN Associates Inc. July 9, 2001.
    <http://www.precarn.ca/IntelligentSystemsSector/ResesarchProjects/research_projects_AthruK.cfm#APACS>

Chapter 10

1. Harry, Mikel, and Schroeder, Richard. Six Sigma: The Breakthrough Management Strategy Revolutionizing the World's Top Corporations. New York: Random House Inc., 2000. P. 214.

2. Webber, Alan M. "New Math for a New Economy." Fast Company. Issue 31. P. 214.

3. Haugeland, John. Artificial Intelligence: The Very Idea. Boston: MIT Press, 1989. P. 224.

4. Bylinsky, Gene. "The Digital Factory." Fortune. November 14, 1994. Pp. 92-110.

5. "Key Drivers for the Future of Manufacturing". Allen-Bradly. March 29, 2001.
   <http://www.ab.com/events/choices/key.html>

6. "Key Drivers for the Future of Manufacturing."

7. "Key Drivers for the Future of Manufacturing."

8. "Special Report: The New Economy." The Economist. May 12, 2001. Pp. 79-81.

9. Webber.

10. Chisholm, Patricia. "Redesigning Work." <u>Maclean's</u>. March 5, 2001. Pp. 34-38.

11. Massing, Dana. "Training Critical to Manufacturing's Future." <u>Erie Times News.</u> March 3, 2001 <http://www.goerie.com/erie2000/training_critical_to_manufactu.html>

12. Massing.

13. Malthus, T. R. <u>On the Principle of Population</u>. Vol. 1. Great Britain: J.M. Dent & Sons Ltd., 1952. Chapter 1, P. 5.

14. IdeMart Design for Sustainability Program, Sub-Faculty of Industrial Design Engineering, Delft University of Technology, Holland, 2000.

15. Mao, Bruce. <u>Life Style</u>. Phaidon Press 2000. P. VI.

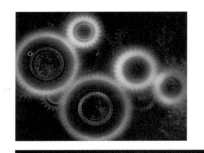

Glossary

Glossary

Note: This glossary is provided to assist the reader in understanding the meaning of technical terms used in this book, in the context of the subject under discussion and is consistent with the reference material. Some of the terms may have more than one definition in other contexts.

Activators: a mechanical or electrical device, which executes the instruction of a control system, by changing a process variable. (p. 60)

Active centers: centers of a process stream, where a transformation of the product physically takes place, for example a milling machine. (p. 42)

Activity-based accounting: a managerial accounting practice that focuses on operational aspects of providing products, services, and projects to customers. (p. 49)

Adaptive behavior: the capability of a machine to learn from the results of a previous operation and change preset algorithms to achieve results closer to the desired target. (p. 63)

Agile manufacturing: a set of practices that extend the capabilities of lean manufacturing and quality systems to the accomplishment of rapid changeover to the manufacture of an alternate product. (p. 164)

Analog systems: systems that represent or process data in continuously variable physical quantities, in contrast to the digital representation and processing of data in discrete units. (p. 69)

Artificial intelligence: the capability of a machine or a network of machines to analyze input and output signals and make independent decisions. See Chapter 4, How Machines Learn. (p. 15)

Assembly processes: a process which creates a product by combining prefabricated components, usually with minimal transformation of the components. For example, assembling a washing machine from parts. (p. 14)

Automatically guided vehicles (AGV): machines designed to provide robotic transport of goods throughout a variety of manufacturing and specialty environments. (p. 66)

Available time: time that is available (usually calendar time less down shifts) for a machine to be operated, whether the machine is running or not. (p. 45)

Blanking: the cutting out of sheet (from rolls/coils or larger pieces) into shapes of useable size. The arrangement of these shapes (nesting) largely determines the yield or recovery. (p. 44)

Bottleneck: that operation in a system where the flow is most constrained. (p. 46)

Capital asset productivity: the value of sales in dollars generated by one dollar of capital employed (= $sales/$capital employed). The inverse ratio (capital employed/sales) is called assets turnover. (p. 164)

Capital deepening: the process of achieving productivity gains by investing more capital in the enterprise. (p. 173)

Capital employed: fixed assets plus current assets minus current liabilities. Capital employed is the value of the assets that contribute to a company's ability to generate revenue. (p. 50)

Common causes of variation: causes of variation attributable to the technical process or business process design. (p. 94)

Communication network: the hardware, software, and architecture of a communication system. (p. 23)

Communications architecture: a definition of the communications connections and methodologies within a system. (p. 23)

Compound yield: the overall recovery of material through a process stream. Calculated as the final part weight divided by the initial part weight, expressed in percent. (p. 53)

Connectivity: the ability to make and maintain a connection between two or more points in a telecommunications system. (p. 172)

Control logic: algorithms which use the information available from the system to adjust the operation policies that affect system behavior, based on a defined objective and a set of the constraints. (p. 61)

Critical input variable (CIV): the variables selected to bring back the output control variables to the desired set point. (p. 95)

Critical output variable (COV): the variables to be measured in order to monitor the system performance or product quality. (p. 95)

Cumulative distribution: the cumulative distribution function for a discrete random variable (X) provides the cumulative probabilities $P(X<x)$ for all values of x. (p. 167)

Cyborg: a mutualistic relation of man and machine. (p. 141)

Cycle time: in this book, cycle time is defined as the time spent in a process or process stream. Sometimes defined elsewhere as the rate of processing (so many parts per minute come off the line). (p. 49)

Decision trees: a methodology for graphically representing decision alternatives with uncertain outcomes. Spreadsheet add-ins can be purchased that make these particularly useful. (p. 155)

Digital systems: systems that employ digital (0 or 1) representation of data in discrete units. See "Analog systems" for the alternative. (p. 70)

Disassembly processes: a process which creates a product by separating parts of an existing product or material. For example, sorting gravel in different grades of sieve size. (p. 62)

Discrete events: events that happen at particular instant in time. (p. 50)

Discrete events simulation: the process of creating a virtual description of an operation by flowing items, called entities, through a series of stations, where the attributes of the entities are changed. The algorithms for doing this are available in commercial software (Arena, Witness, Simul8, and several others) and are often an extension of CAD. The software does calculations only when an event occurs at a particular instant in time. (p. 50)

Distributed information systems: information systems (that aspect of an organization or organizations that is concerned with the storage, transfer, and processing of information) that consist of multiple independently functioning but interacting parts. (p 54)

Dynamic modeling (simulation): dynamic simulation model represent a system as it evolves over time. (p. 55)

Engineered scrap: the scrap resulting from intended processing. For example, trimming a part will result in some engineered scrap. (p. 44)

Enterprise resource planning (ERP): An amalgamation of a company's information systems designed to bind more closely a variety of company functions including human resources, inventories and financials while simultaneously linking the company to customers and vendors. (p. 61)

Error Analyzing System (EAS): an automatic process, common in intelligent machines, which analyzes statistically the variation of measurements from a sensor with reference to a known target, and applies a recalibration if the variation is outside the control limits. (p. 97)

EVA: "Economic Value Added." For a company, after-tax earnings minus the opportunity cost of capital (NOPAT - capital charge). This essentially measures how much more valuable a company has become during a given time period. (p. 44)

Expert system: a collection of empirical rules, often in the form of a database, which represent the best experience with a process or series of processes. Intelligent machines use this as one of the references for adaptation. Expert systems are embodied in specialized software. (p. 73)

Feed-forward: a control strategy where critical input variable behavior is modeled or empirically measured and the process is preset to the expected behavior. This reduces the effect of disturbances before they have a chance to affect a critical output variable. (p. 64)

Feedback: a control strategy, where the process critical output variable to be controlled is measured and this measurement is used to adjust critical input variables. (p. 65)

Finite-element analysis: a computer-based numerical technique for calculating the strength and behavior of engineering structures. In the finite-element method, a structure is broken down into many small simple blocks or elements. The behavior of an individual element can be described with a relatively simple set of equations. Just as the set of elements would be joined together to build the whole structure, the equations describing the behaviors of the individual elements are joined into an extremely large set of equations that describe the behavior of the whole structure. (p. 149)

Fishbone diagram: also known as cause and effect diagram or Ishikawa diagram after its inventor Dr. K. Ishikawa, this is a diagram composed of lines and symbols designed to represent a meaningful relationship between an effect and its causes. (p. 95)

Fixed cost: a cost that does not vary depending on production or sales levels, such as rent, property tax, insurance, or interest expense. (p. 54)

Functional organization: An organization based on grouping people with similar functions into separate departments (usually hierarchically). (p. 127)

GAAP: "Generally Accepted Accounting Principles." A widely accepted set of rules, conventions, standards, and procedures for reporting financial information, as established by an accounting regulating body in each country or group of countries. (p 49)

GDP: "Gross Domestic Product." The total market value of all final goods and services produced in a country in a given year; equals total consumer, investment and government spending, plus the value of exports minus the value of imports. (p. 172)

Genetic algorithms: a biologically inspired search method that seeks to converge to a solution using an evolutionary process. It is used when other direct optimization algorithms are not effective. (p. 76)

Heuristic data: data resulting from a problem-solving technique in which the most appropriate solution of several found by alternative methods is selected at successive stages of a program for use in the next step of the program. (p. 158)

Holon: an autonomous, intelligent, distributed, cooperative device. It could be a robot, a PLC, a person, or any other component of a process. (p. 172)

HQP: "Highly Qualified Professionals." A term attributed to people with a high level of education acquired through academic achievement, postgraduate studies, and industrial experience. (p. 175)

Indirect cost: the cost not directly attributable to the manufacturing of a product. Opposite of direct cost. (p. 54)

Intelligent machines: machines that have the capacity to communicate with their operators and each other, and sometimes even the capacity to learn. (p. 58)

LAN: "Local Area Network." A data communications network, which is geographically limited (typically to a 1 km radius) allowing easy interconnection of terminals, microprocessors, and computers within adjacent buildings.

LCA: "Life-cycle analysis." A standardized methodology for measuring the total environmental impact of producing, using, and disposing of a product. (p. 17)

Linear or arithmetic progression: a simple form of sequence or series relationship with the general form $Xn = a + nd$ where a is the initial value, n is the position of the number X in the series, and d is the rate of progression. The plot is a straight line with intercept a on the Y-axis and slope d.

Linear programming: a mathematical technique for solving optimization problems in which all the constraints and objective functions are linear functions of a number of real variables. (p. 23)

Machine availability: the long-term fraction of the time that the machine is ready to process parts.

Machine reliability: the probability that a machine will perform a required function for a given period of time when used under stated operating conditions. (p. 106)

Mass customization: the process of creating individually tailored products by the assembly of mass-produced standard components. (p. 8)

Mathematical modeling: models constructed using a set of mathematical equations or logical relationships to describe the real system. (p. 121)

Matrix organization: a combination of functional organization and product organization requiring parallel hierarchical structures. (p. 133)

Mean: quantity which has a value intermediate between the extreme numbers of a set. Several kinds of means exist and the method of calculating them depends on the relationship, known or assumed, to the other numbers of the set. The arithmetical mean, used in this book represents a point around which the numbers balance $(= S_n /n)$. (p. 83)

Median: a value which divides the set of observations into two groups so that the number of observations above it is equal to the number below it. (p. 83)

Mini-mill: a small mill or plant, especially a steel or aluminium mill that uses continuous processes and limited infrastructure. (p. 147)

Mode: the value with the highest frequency of numbers in a set. (p. 83)

NC machines: "Numerically Controlled Machines." Machines capable of receiving instructions in the form of a program (tape or disk or program download from a network) and executing it. (p. 64)

NEBIT: "Net Earnings Before Interest and Taxes." Same as operating income.

Nested blanking: blanking process where shapes to be cut are arranged to optimize (or nearly) yield. (p. 44)

Neural network: a system of programs and data structures that approximates the operation of the human brain. (p. 74)

NGM: "Next Generation Manufacturing." A U.S. government-industry-academia study commissioned in the early 1990s to create a vision of manufacturing for the next century. (p. 145)

Nonactive centers: centers in a product stream where business process activities take place, which are not related to the transformation of the product. For example, warehousing, production planning, accounting. (p. 43)

NOPAT: "Net Operating Profit After Taxes" = operating income x $(1 - \text{tax rate})$. A profitability measure that omits the cost of debt financing (i.e., it omits interest payments, along with their associated tax break). NOPAT is primarily used in the calculation of EVA.

Normal distribution or Gaussian distribution: a frequency distribution of variables showing their observed or theoretical frequency of occurrence with a symmetrical occurrence of values on either side of the mean.

Normal margin: in this book, defined as the difference between the normal value of cost and the normal value of price. (p. 108)

Off-line quality control: a method of controlling quality by taking statistically significant samples of finished products and analyzing them in a test station or laboratory, usually operated by a quality control department. This is the traditional method used by manufacturing and is being replaced by statistical process control. (p. 146)

Open systems: systems where the details exist in the public domain. (p. 172)

Open-book management: a management style, which allows all employees to access strategic goals and performance indicators of the business that are usually the prerogative of upper management. (p. 131)

Optimization algorithms: algorithms that seek the most advantageous value of a specified variable. (p. 48)

Order entry: the point in the product stream where customer orders are received by the firm and placed into the system as work items. (p. 61)

Organic computers: computer technology inspired by ideas and materials from biology, not necessarily reliant on binary logic. (p. 72)

Pattern recognition: a field that applies various concepts, tools, and techniques to place an unknown pattern into one of a set of predefined classes of patterns. (p. 76)

Payroll: the financial record of employees' salaries, wages, bonuses, net pay, and deductions. (p. 159)

Planned time (also called *scheduled time*): the time that a plant or equipment is scheduled to be operational. (p. 45)

PLC: "Programmable Logic Controller." A device used to automate monitoring and control of industrial plant and/or equipment. *Population:* a group of all possible elements that could be measured or observed. (p. 76)

Population: a group of all possible elements that could be measured or observed. (p.76)

Process control chart: a graphical method used to show the results of a small-scale repeated sampling of a manufacturing process. (p. 78)

Process in control: also referred to as a "stable" process. A process in which all the assignable causes are removed and the averages and variations of the desired quality characteristics of the samples are within control limits. (p. 88)

Product cost accounting: the process of identifying and evaluating production costs on a per-product basis. (p. 51)

Product-oriented organization: an organizational structure where the duties and responsibilities of employees are grouped by product divisions. (p. 86)

Product stream: the manufacturing segment of the supply chain for a particular product. (p. 43)

Production schedule: the allocation of system resources in order to manage on-time delivery, minimal work in process, short customer lead times, and maximum utilization of resources. (p. 61)

Productivity: There are many definitions of productivity depending on the denominator of the equation "production/y." For example if y is time, productivity is a measure of output (mass flow). If y is hours of labor, productivity is a measure of the effectiveness of the workers. Usually these measures are relative to a chosen standard and productivity is measured as a variance (percentage increase or decrease). (p. 168)

Programmed sequences: serial operations, which are computer controlled according to predetermined sequences. (p. 60)

Pull system: a system that enters parts into the system based on the real or predicted demand. (p. 114)

Push system: a system that authorizes production as inventory is consumed, i.e., Kanban system. (p. 114)

Qualitative product attributes: those attributes of a product that are not strictly measurable, such as perceived quality, or utility. (p. 117)

Quantitative product attributes: those attributes of a product that are measurable, such as part mass, or size. (p. 117)

Queues: the line-ups of parts that form in front of a constraining process step, or in front of batching processes. (p. 61)

Real time: time is a continuum between past, present, and future. In automation terminology, real time means the present. Real-time data are read directly from machines as the process happens. (p. 15)

Rheometer: an instrument for measuring the flow of viscous liquids such as polymers or blood. (p. 149)

Risk analysis: a structured methodology to evaluate and quantify the outcomes in an uncertain environment. (p. 155)

Scheduling: the process of determining what should be manufactured when in an industrial plant. (p. 166)

Self-tending machines: automated machines capable of monitoring their own performance and performing recalibration and repairs with no human assistance. (p. 125)

Sensors: instruments that can measure system output variables. (p. 60)

Setup time: time required to readjust a machine in order to make it ready to process a different product type. (p. 106)

Six Sigma: a comprehensive system for building and sustaining business performance, success, and leadership with the goal of near-perfection (not more than 3.4 defects per million transactions). (p. 105)

Slide ruler: a device consisting of two logarithmically scaled rulers mounted to slide along each other so that multiplication, division, and other more complex computations are reduced to the mechanical equivalent of addition or subtraction. (p.70)

Special causes of variation: causes of variation attributable to errors in the operation of the technical or business process. (p. 94)

Standard deviation: a measure of dispersion of a set of data. It is equal to the square root of its variation. (p. 93)

Standard normal distribution: a normal distribution with mean zero and standard deviation of one. A normal distribution can result from the original data values being expressed in terms of their number of standard deviations away from the mean. (p. 106)

Static simulation model: a representation of a system at a particular time or a model where none of its inputs or outputs is a function of time. (p. 104)

Statistical process control: a method for measuring, understanding, and controlling variation in a manufacturing process. (p. 41)

Steady state: a state of equilibrium when a reference output of a system is constant and not a function of time. (p. 64)

Stream cycle time or time-in-system: the time spent in a process stream by an individual part or entity being processed. (p. 49)

Swaging: bending or shaping cold metal through the use of a die. (p. 168)

System dynamics simulation: the process of creating a virtual description of an operation by flowing items through flow-control logic blocks and accumulators. The algorithms for doing this are available in commercial software (Ithink, GoldSim, and several others). The software does calculations, which verify the state of the system, at fixed intervals of time, regardless of the activities taking place. (p. 84)

Tailor-welded blanks: blanks that are used in stamping operations that have different sicknesses of sheet welded onto critical areas to provide more strength. The welding is performed prior to stamping, resulting in a significant weight and material savings when compared to using a uniform sheet thickness throughout the part. (p. 44)

Technical-gap measurement: a measure of the gap between what is being done at present within the system, and what is technically achievable. This gap can be measured for productivity, quality, or other quantifiable metric. (p. 149)

Tele-operator: a machine/human interaction where a complex task is divided between the human operator and the machine. (p. 68)

Throughput: may also be referred to as "throughput rate." The average of the output of a production process such as a machine, combinations of machines, or even the whole system, per unit of time. (p. 101)

Throughput accounting: a management accounting system based on the Theory of Constraints (TOC). Throughput accounting does not allocate costs to products, but only to the system as a whole. (p. 101)

Time-in-system: see Stream cycle time. (p. 49)

Transformation processes: processes where there is a physical transformation of the entity flowing through the system; for example, a melting operation is a transformation process, whereas storing inventory in a warehouse is not. (p. 62)

Transients: conditions that change over time, or that only exist for a certain time. (p. 64)

Utility analysis: a methodology that incorporates a firm or person's attitude toward risk into an economic decision analysis. (p. 155)

Value pricing: in highway tolling, also known as congestion pricing and peak-period pricing. A way of harnessing the power of the market and reducing the waste associated with congestion. It entails fees or tolls for road use, which vary with the level of congestion. In this book we have used the broader definition by Erwin Gross of the limit that a customer is willing to pay for a given functionality. (p. 102)

Variable cost: cost that necessarily varies with production volume. (p. 49)

Variance: a measure of scattering around the zone of the central tendency based on the square differences between observed values and the mean. (p. 37)

Vectors: a measure that has both a scalar metric and a direction. (p. 77)

Virtual environment: an artificial environment created through computer programming, which allows interactive exchange with the user through sensors and activators (for example, a flight simulator). (p. 141)

Virtual factory: a manufacturing system consisting of physically dispersed operations, often hundreds or thousands of miles distant, connected by networks and specialized computer software. (p. 58)

Virtual operations: manufacturing steps which simulate a physical connection (like a conveyor or a pipe) between two separate machines. Virtual operations use real-time information from machines and from human planners to activate the next step in a process stream with a minimum delay. (p. 53)

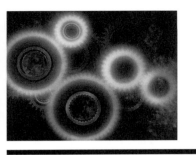

Index

A

Activators: 60

Active centers: 42

Activity based accounting: 49

Adaptive behavior: 63

Agile manufacturing: 164

Analog systems: 69

Artificial intelligence: 15

Assembly processes: 14

Automatically guided vehicles (AGV): 66

Available time: 45

B

Blanking: 44

Bottleneck: 46

C

Capital asset productivity: 164

Capital deepening: 173

Capital employed: 50

Common causes of variation: 94

Communication network: 23

Communications architecture: 23

Compound yield: 53

Connectivity: 172

Control logic: 61

Critical input variable (CIV): 95

Critical output variable (COV): 95

Cumulative distribution: 167

Cyborg: 141

Cycle time: 49

D

Decision trees: 155

Digital systems: 70

Disassembly processes: 62

Discrete events: 50

Discrete events simulation: 50

Distributed information systems: 54

Dynamic modeling (simulation): 55

E

Engineered scrap: 44

Enterprise resource planning (ERP): 61

Error analyzing system (EAS): 97

EVA: 44

Expert system: 73

F

Feed-forward: 64

Feedback: 65

Finite-element analysis: 149

Fishbone diagram: 95

Fixed cost: 54

Functional organization: 127

G

GAAP: 49

GDP: 172

Genetic algorithms: 76

H

Heuristic data: 158

Highly Qualified People: 175

Holon: 172

I

Indirect cost: 54

Intelligent machines: 58

L

LCA (life-cycle analysis): 17

Linear programming: 23

M

Machine reliability: 106

Mass customization: 8

Mathematical modeling: 121

Matrix organization: 133

Mean: 83

Median: 83

Mini-mill: 147

Mode: 83

N

Nested blanking: 44

Neural network: 74

Next Generation Manufacturing: 145

Nonactive centers: 43

Normal margin: 108

Numerically controlled machines: 64

O

Off-line quality control: 149

Open systems: 172

Open-book management: 131

Optimization algorithms: 48

Order entry: 61

Organic computers: 72

P

Pattern recognition: 76

Payroll: 159

Planned time/scheduled time: 45

Population: 76

Process control chart: 78

Process in control: 88

Product cost accounting: 51

Product-oriented organization: 86

Product stream: 43

Production schedule: 61

Productivity: 168

Programmed sequences: 60

Pull system: 114

Push system: 114

Q

Qualitative product attributes: 117

Quantitative product attributes: 117

Queues: 61

R

Real time: 15

Rheometer: 149

Risk analysis: 155

S

Scheduling: 166

Self-tending machines: 125

Sensors: 60

Setup time: 106

Six Sigma: 105

Slide ruler: 70

Special causes of variation: 94

Standard deviation: 93

Standard normal distribution: 106

Static simulation model: 104

Statistical process control: 41

Steady state: 64

Stream cycle time or time-in-system: 49

Swaging: 168

System dynamics simulation: 84

T

Tailor-welded blanks: 44

Technical-gap-measurement: 149

Tele-operator: 68

Time-in-system: 49

Throughput: 101

Throughput accounting: 101

Transformation processes: 62

Transients: 64

U

Utility analysis: 155

V

Value pricing: 102

Variable cost: 49

Variance: 37

Vectors: 77

Virtual factory: 58

Virtual environment: 141

Virtual operations: 53